T0214493

Communications
in Computer and Information Science   1040

Commenced Publication in 2007
Founding and Former Series Editors:
Phoebe Chen, Alfredo Cuzzocrea, Xiaoyong Du, Orhun Kara, Ting Liu,
Krishna M. Sivalingam, Dominik Ślęzak, Takashi Washio, and Xiaokang Yang

More information about this series at http://www.springer.com/series/7899

Dimitris Kotzinos · Dominique Laurent ·
Nicolas Spyratos · Yuzuru Tanaka ·
Rin-ichiro Taniguchi (Eds.)

Information Search, Integration, and Personalization

12th International Workshop, ISIP 2018
Fukuoka, Japan, May 14–15, 2018
Revised Selected Papers

 Springer

Editors
Dimitris Kotzinos
Lab. ETIS UMR 8051
University of Paris-Seine, University
of Cergy-Pontoise, ENSEA, CNRS
Cergy-Pontoise, France

Nicolas Spyratos
LRI
University of Paris-Sud
Orsay, France

Rin-ichiro Taniguchi
Kyushu University
Fukuoka, Japan

Dominique Laurent
Lab. ETIS UMR 8051
University of Paris-Seine, University
of Cergy-Pontoise, ENSEA, CNRS
Cergy-Pontoise, France

Yuzuru Tanaka
Hokkaido University
Sapporo, Japan

ISSN 1865-0929   ISSN 1865-0937 (electronic)
Communications in Computer and Information Science
ISBN 978-3-030-30283-2   ISBN 978-3-030-30284-9 (eBook)
https://doi.org/10.1007/978-3-030-30284-9

This Springer imprint is published by the registered company Springer Nature Switzerland AG
The registered company address is: Gewerbestrasse 11, 6330 Cham, Switzerland

Preface

This book contains the selected research papers presented at ISIP 2018, the 12th International Workshop on Information Search, Integration and Personalization, held during May 14–15, 2018, at Kyushu University, Fukuoka, Japan.

Two keynote talks (whose abstracts are included here) were given during the workshop:

– Professor Makoto Yokoo (Kyushu University, Japan), "Market Design for Constrained Matching"
– Professor Sri Parameswaran (University of New South Wales, Australia), "Side Channel Attacks in Embedded Systems: A Tale of Hostilities and Deterrence"

There were 32 presentations of scientific papers, of which 13 were submitted to the post-workshop peer review. The international Program Committee selected seven papers to be included in the proceedings. The committee decided to also include an invited paper:

– Michel De Rougemont and Guillaume Vimont (University Paris II, France), "Approximate Integration of Streaming Data"

This invited paper proposes an approach to integrating data streams coming from data warehouse or social networks or a combination thereof. Given the big volumes of data involved, the approach proposes approximation techniques for answering queries using a weighted reservoir sampling. In the case of graph edges from a social network, the communities are approximated as the large connected components of the edges in the reservoir. It is shown that for a model of random graphs which follow a power law degree distribution, the community detection algorithm is a good approximation. Given two streams of graph edges from two sources, the content correlation of the two streams is defined to the fraction of the nodes in common communities in both streams. Although the edges of the streams are not stored, the authors approximate the content correlation online and define the integration of two streams as the collection of the large communities. The approach is illustrated with Twitter streams.

The themes of the presented and/or submitted papers reflected today's diversity of research topics as well as the rapid development of interdisciplinary research. With increasingly sophisticated research in science and technology, there is a growing need for interdisciplinary and international availability, distribution, and exchange of the latest research results, in organic forms, including not only research papers and multimedia documents, but also various tools developed for measurement, analysis, inference, design, planning, simulation, and production as well as the related large data sets. Similar needs are also growing for the interdisciplinary and international availability, distribution, and exchange of ideas and works among artists, musicians, designers, architects, directors, and producers. These contents, including multimedia documents, application tools, and services are being accumulated on the Web, as well as in local and global databases, in a remarkable speed that we have never experienced with other kinds of publishing media. Large amounts of content are now already on the Web, waiting for their advanced personal and/or public reuse. We need new theories and technologies for the advanced information search, integration through interoperation, and personalization of Web content as well as database content.

The ISIP 2018 workshop was organized to offer a forum for presenting original work and stimulating discussions and exchanges of ideas around these themes, focusing on the following topics.

- Data Quality
- Social Cyber-Physical Systems
- Information search in large data sets (databases, digital libraries, data warehouses)
- Comparison of different information search technologies, approaches, and algorithms
- Novel approaches to information search
- Personalized information retrieval and personalized Web search
- (Big) Data Analytics
- Integration of Web-services, Knowledge bases, Digital libraries
- Federation of Smart Objects
- Machine learning and AI
- Visual and sensory information processing and analysis

The selected papers contained in this book are grouped into three major topics, namely Data Integration, Text and Document Management, and Advanced Data Mining Techniques; they span major topics in Information Management research both modern and traditional.

Historical Note

ISIP started as a series of Franco-Japanese workshops in 2003, and its first edition took place under the auspices of the French embassy in Tokyo, which provided the financial support along with JSPS (Japanese Society for the Promotion of Science). Up until 2012, the workshops have alternated between Japan and France, and attracted increasing interest from both countries. Then, motivated by the success of the first editions of the workshop, participants from countries other than France or Japan volunteered to organize it in their home country.

The history of past ISIP workshops is as follows:

- 2003: 1st ISIP in Sapporo (June 30 to July 2, Meme Media Lab, Hokkaido University, Japan)
- 2005: 2nd ISIP in Lyon (May 9–11, University Lyon 1, France)
- 2007: 3rd ISIP in Sapporo (June 27–30, Meme Media Laboratory, Hokkaido University, Japan)
- 2008: 4th ISIP in Paris (October 6–8, Tour Montparnasse, Paris, France)
- 2009: 5th ISIP in Sapporo (July 6–8, Meme Media Laboratory, Hokkaido University, Japan)

- 2010: 6th ISIP in Lyon (October 11–13, University Lyon 1, France)
- 2012: 7th ISIP in Sapporo (October 11–13, Meme Media Laboratory, Hokkaido University, Japan)
- 2013: 8th ISIP in Bangkok (September 16–18, Centara Grand and Bangkok Convention Centre CentralWorld Bangkok, Thailand).
- 2014: 9th ISIP in Kuala Lumpur (October 9–10, HELP University, Kuala Lumpur, Malaysia)
- 2015: 10th ISIP in Grand Forks (October 1–2, University of North Dakota, Grand Forks, North Dakota, USA)
- 2016: 11th ISIP in Lyon (November 3–4, University Lyon 1, France)

Originally, the workshops were intended for a Franco-Japanese audience, with the occasional invitation of researchers from other countries as keynote speakers. The proceedings of each workshop were published informally, as a technical report of the hosting institution. One exception was the 2005 workshop, selected papers of which were published by the *Journal of Intelligent Information Systems* in its special issue for ISIP 2005 (Vol. 31, Number 2, October 2008). The original goal of the ISIP workshop series was to create close synergies between a selected group of researchers from the two countries; and indeed, several collaborations, joint publications, joint student supervisions, and research projects originated from participants of the workshop.

After the first six workshops, the organizers concluded that the workshop series had reached a mature state with an increasing number of researchers participating every year. As a result, the organizers decided to open up the workshop to a larger audience by inviting speakers from over ten countries at ISIP 2012, ISIP 2013, ISIP 2014 as well as at ISIP 2015. The effort to attract an even larger international audience has led to the workshop being organized in countries other than France and Japan. This will continue in the years to come. During the last six years in particular, an extensive effort was made to include in the Program Committee academics coming from around the globe, giving the workshop an even more international character.

We would like to express our appreciation to all the staff members of the organizing institution for the help, kindness, and support before, during, and after the workshop. Of course we also would like to cordially thank all speakers and participants of ISIP 2018 for their intensive discussions and exchange of new ideas. This book is an outcome of those discussions and exchanged ideas. Our thanks also go to the Program Committee members whose work has been undoubtedly essential for the selection of the papers contained in this book.

April 2019

Nicolas Spyratos
Yuzuru Tanaka
Rin-Ichiro Taniguchi
Co-chairs of ISIP 2018

Organization

ISIP 2018 was organized by Kyushu University, in Fukuoka, Japan

Executive Committee

Co-chairs

Nicolas Spyratos	Paris-Sud University, France
Yuzuru Tanaka	Hokkaido University, Japan
Rin-Ichiro Taniguchi	Kyushu University, Japan

Program Committee Chairs

Dimitris Kotzinos	University of Cergy-Pontoise, France
Dominique Laurent	University of Cergy-Pontoise, France

Local Organization

Maiya Hori	Kyushu University, Japan
Koji Okamura	Kyushu University, Japan
Kenji Ono	Kyushu University, Japan
Atsushi Shimada	Kyushu University, Japan
Hideaki Uchiyama	Kyushu University, Japan

Publicity Chair

Diego Thomas	Kyushu University, Japan

Program Committee

Giacometti, Arnaud	Université François Rabelais de Tours, France
Jen, Tao-Yuan	University of Cergy-Pontoise, France
Kotzinos, Dimitris	University of Cergy-Pontoise, France
Laurent, Anne	Université Montpellier, France
Laurent, Dominique	University of Cergy-Pontoise, France
Lhouari, Nourine	Université Clermont Auvergne, France
Lucchese, Claudio	Ca' Foscari University of Venice, Italy
Okada, Yoshihbiro	Kyushu University, Japan
d'Orazio, Laurent	Université de Rennes 1, France
Petit, Jean-Marc	INSA de Lyon, France
Plexousakis, Dimitris	FORTH Institute of Computer Science, Greece
de Rougemont, Michel	Universit Paris II, France
Sais, Lakhdar	Université d'Artois, France
Spyratos, Nicolas	Université Paris-Sud, France
Sugibuchi, Tsuyoshi	CustomerMatrix Inc., France

Tanaka, Yuzuru   Hokkaido University, Japan
Taniguchi, Rin-Ichiro   Kyushu University, Japan
Vodislav, Dan   University of Cergy-Pontoise, France
Yoshioka, Masaharu   Hokkaido University, Japan

Abstracts of Keynote Talks

Market Design for Constrained Matching

Makoto Yokoo

Kyushu University, Japan

The theory of two-sided matching (e.g., assigning residents to hospitals, students to schools) has been extensively developed, and it has been applied to design clearing-house mechanisms in various markets in practice. As the theory has been applied to increasingly diverse types of environments, however, researchers and practitioners have encountered various forms of distributional constraints. As these features have been precluded from consideration until recently, they pose new challenges for market designers. One example of such distributional constraints is a minimum quota, e.g., school districts may need at least a certain number of students in each school in order for the school to operate. In this talk, I present an overview of research on designing markets/mechanisms that work under distributional constraints.

Side Channel Attacks in Embedded Systems: A Tale of Hostilities and Deterrence

Sri Parameswaran

University of New South Wales, Australia

Deep devastation is felt when privacy is breached, personal information is lost, or property is stolen. Now imagine when all of this happens at once, and the victim is unaware of its occurrence until much later. This is the reality, as increasing amount of electronic devices are used as keys, wallets and files. Security attacks targeting embedded systems illegally gain access to information or destroy information. Advanced Encryption Standard (AES) is used to protect many of these embedded systems. While mathematically shown to be quite secure, it is now well known that AES circuits and software implementations are vulnerable to side channel attacks. Side-channel attacks are performed by observing properties of the system (such as power consumption, electromagnetic emission, etc.) while the system performs cryptographic operations. In this talk, differing power based attacks are described, and various countermeasures are explained. In particular, a countermeasure titled Algorithmic Balancing is described in detail. Implementation of this countermeasure in hardware and software is described. Since process variation impairs countermeasures, we show how this countermeasure can be made to overcome process variations.

Contents

Data Integration

Approximate Integration of Streaming Data

Michel De Rougemont[1,2(✉)] and Guillaume Vimont[1,2]

[1] IRIF-CNRS, Paris, France
mdr@irif.fr
[2] Univ. Paris II, Paris, France

Abstract. We study streams of tuples of a Datawarehouse and streams of edges of a dynamic graph. In the case of a Datawarehouse we show how to approximate some OLAP queries with a weighted Reservoir sampling. In the case of graph edges from a Social Network, we approximate the communities as the large connected components of the edges in the Reservoir. We show that for a model of random graphs which follow a power law degree distribution, the community detection algorithm is a good approximation. Given two streams of graph edges from two sources, we define the *Content Correlation* of the two streams as the fraction of the nodes in common communities in both streams. Although we do not store the edges of the streams, we approximate the Content Correlation online and define the *Integration of two streams* as the collection of the large communities. We illustrate this approach with Twitter streams.

Keywords: Streaming algorithms · Community detection ·
Dynamic random graphs · OLAP queries

1 Introduction

The integration of several sources of data is also called the composition problem, in particular when the sources do not follow the same schema. It is relevant for two distinct Datawarehouses, two Social networks, or one Social Network and one Datawarehouse. We specifically study the case of two streams of labeled graphs from a Social Network and develop several tools using randomized streaming algorithms. We define the *Content Correlation* between two streaming graphs built from sequences of edges and study how to approximate it with an online algorithm.

The basis of our approach is the approximation of analytical queries, in particular when we deal with streaming data. In the case of a Datawarehouse, we have a stream of tuples t following an OLAP schema, where each tuple has a measure, and we want to approximate OLAP queries. In the case of a Social network such as Twitter, we have a stream of tweets which generate edges of an evolving graph, and we want to approximate the evolution of the communities as a function of time.

© Springer Nature Switzerland AG 2019
D. Kotzinos et al. (Eds.): ISIP 2018, CCIS 1040, pp. 3–22, 2019.
https://doi.org/10.1007/978-3-030-30284-9_1

The main randomized technique used is a *k-weighted Reservoir sampling* which maps an arbitrarly large stream of tuples t of a Datawarehouse to k tuples whose weight is the measure $t.M$ of the tuple. It also maps a stream of edges $e = (u, v)$ of a graph, to k edges and in this case the measure of the edges is 1. We show how we can approximate some OLAP queries and the main study will be the approximate dynamic community detection for graphs, using only a dynamic Reservoir. We do not store the edges of the graph and at any given time we only maintain a Reservoir with k random edges and compute the connected components of these edges. We interpret the large connected components as communities and follow their evolution in time.

The edges of the Reservoir are taken with a uniform distribution over the edges, hence the nodes of the edges are taken with a probability proportional to their degrees. In a sliding window, a Reservoir can be implemented with techniques such as Minhash [5] which generalizes the Reservoir sampling [16]. At any given time t, we have a Reservoir which keeps k edges among the possible m edges which have been read ($m \gg k$), and each edge has the same probability k/m to be chosen in the Reservoir. If a cluster S is large enough, it will show in the Reservoir as a large connected component, because the random edges of S are taken with the same probability $p = k/m$, i.e. follow the Erdös-Renyi model $G(n, p)$ and the giant component in S occurs if $p > 1/n = 1/|S|$, the classical phase transition probability. This simple approach scales for dynamic streams. We will observe changes in the communities as new communities appear and old communities disappear. At discrete times t_i we store only the large connected components of the Reservoirs.

We introduce the *Content Correlation* ρ between two streams as the proportion of nodes at the intersection of the communities of each stream. We show how to approximate the correlation with an online algorithm. The integration of two streams of edges defining two graphs $G_i = (V_i, E_i)$ for $i = 1, 2$ can then be viewed as the new structure which collects the large connected components in the Reservoirs of both streams.

Our main application is the analysis of Twitter streams: a stream of graph edges for which we apply our k-Reservoir. We temporarily store the k-edges and only store the large connected components, i.e. of size greater than some constant h and their evolution in time. We give examples from the analysis of streams associated with TV channels and crypto-currencies and their correlation. Our main results are:

- An approximation algorithm for the community detection for graphs following a power law degree distribution with a concentrated selection (Theorem 1), its extension to dynamic graphs (Theorem 2).
- An online algorithm to compute the correlation $\rho(t)$ and its application to two Steps Dynamics (Theorem 3),
- A concrete analysis on Twitter streams to illustrate the model, and the community correlation of Twitter streams.

We review the main concepts in Sect. 2. We study the approximation of OLAP queries in a stream in Sect. 3. In Sect. 4, we consider streams of edges in a graph and give an approximate algorithm for the detection of communities. In Sect. 5, we define the integration of streams and explain our experiments in Sect. 6.

2   Preliminaries

We introduce our notations for OLAP queries, Social Networks, and the notion of approximation used.

2.1   Datawarehouses and OLAP Queries

A Datawarehouse I is a large table storing tuples t with attributes $A_1, ... A_m, M$, some A_i being foreign keys to other tables, and M a measure. Auxiliary tables provide additional attributes for the foreign keys. An OLAP or star schema is a tree where each node is labeled by a set of attributes and the root is labeled by the set of all the attributes of t. An edge exists if there is a functional dependency between the attributes of the origin node and the attributes of the extremity node. The *measure* is a specific node at depth 1 from the root. An OLAP query for a schema S is determined by: a filter condition, a *measure*, the selection of dimensions or classifiers, $C_1, ... C_p$ where each C_i is a node of the schema S, and an aggregation operator (COUNT, SUM, AVG, ...).

A filter selects a subset of the tuples of the Datawarehouse, and we assume for simplicity that SUM is the Aggregation Operator. The answer to an OLAP query is a multidimensional array, along the dimensions $C_1, ... C_p$ and the *measure* M. Each tuple $c_1, ..., c_p, m_i$ of the answer where $c_i \in C_i$ is such that $m_i = \frac{\sum_{t:t.C_1=c_1,...t.C_p=c_p} t.M}{\sum_{t \in I} t.M}$. The absolute measure is $\sum_{t:t.C_1=c_1,...t.C_p=c_p} t.M$ and the relative measure is just normalized so that the sum of the relative measures is always 1. We consider relative measures as answers to OLAP queries Q_C^I in order to view an answer Q as a distribution or *density vector*, on dimension C and on data warehouse I, as in Fig. 2.

Example 1. Consider tuples $t(ID, Tags, RT, Time, User, SA)$ storing some information about Twitter tweets. Let *Content* = {Tags, RT} where Tags is the set of tags of the Tweet and RT = 1 if the tweet is a ReTweet and RT = 0 otherwise.

The Sentiment Analysis of a tweet is a subjective measure, whether it is positive or negative. For a tuple t, let $t.SA$ be an integer value in $[1, 2, ...10]$. The sentiment is negative if $SA < 5$ and positive when $SA \geq 5$ with a maximum of 10. The simple OLAP schema of Fig. 1 describes the possible dimensions and the measure SA. The edges indicate a functional dependency between sets of attributes.

The dimension $C = Channel$ has two possible values c in the set {CNN, PBS}. The *Analyis per Channel with measure SA* is a distribution Q_C with $Q_{C=CNN}^I = 2/3$ and $Q_{C=PBS}^I = 1/3$ as in Fig. 2. The approximation of Q_C is studied in Sect. 3.

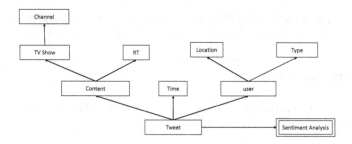

Fig. 1. An OLAP schema for a Datawarehouse storing tuples t for each Twitter tweet, with *Sentiment Analysis*, an integer in $[1, 2, ...10]$ as a measure.

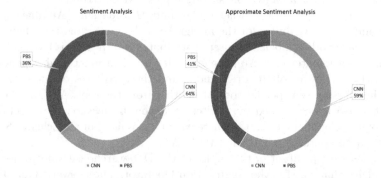

Fig. 2. An OLAP query for the sentiment analysis per channel. The exact solution $Q^I_{C=CNN} = 0.66$ and the approximate solution $Q^I_{C=CNN} = 0.61$ with a Reservoir.

2.2   Social Networks

A social network is a labeled graph $G = (V, E)$ with domain V and edges $E \subseteq V \times V$. In many cases, it is built as a stream of edges $e_1, ...e_m$ wich define E. Given a set of tags, Twitter provides a stream of tweets represented as Json trees. The *Twitter Graph* of a stream is the graph $G = (V, E)$ with multiple edges E where V is the set of tags (#x or @y) seen and for each tweet sent by @y which contains tags #x, @z we add the edges (@y, #x) and (@y, @z) to E. The URL's which appear in the tweet can also be considered as nodes but we ignore them for simplicity.

Social Networks graphs such as the Twiter graph have a specific structure. The graphs are connected, the degree distribution of the nodes follows a power law and the *communities* or clusters are large dense subgraphs. Let $\alpha \leq 1$ and let $E(S)$ be the set of internal edges i.e. edges $e = (u, v)$ where $u, v \in S$. An α-cluster is a subset S such that $|E(S)| \geq \alpha.|S|^2$. We are only interested in large clusters and say that a class of graphs has an α-cluster if there exists α such that for n large enough the graphs have an α-cluster.

The detection of communities is a classical problem, which can be solved by many techniques such as Mincuts, hierarchical clustering or the Girwan-Newman algorithm based on the edge connectivity. All these methods require to store the whole set of edges. By contrast, we will detect communities without storing the entire set of edges, but only samples from the stream of edges, and approximate the dynamics of the communities. We will introduce a technique which integrates two streams.

2.3   Approximation

In our context, we approximate the density values of the OLAP queries or the communities of a graph. We use randomized algorithms with an additive approximation where the probabilistic space Ω for a stream s of m tuples (resp. edges) is the set of k tuples (resp. edges) where each edge occurs with some probability p. In the case of edges, the probability p is uniform, i.e. $p = k/m$. There are usually two parameters $0 \leq \varepsilon, \delta \leq 1$ for the approximation of randomized algorithms, where ε is the error, and $1 - \delta$ the confidence.

In the case of a density value, i.e. a function $F : \Sigma^* \to \mathbb{R}$ where Σ is the set of possible tuples, let A be a randomized algorithm with input s and output $y = A(s)$ where $y \in \mathbb{R}$ is the density value. The algorithm A will (ϵ, δ)-approximate the function F if for all s,

$$Prob_\Omega[F(s) - \varepsilon \leq A(s) \leq F(s) + \varepsilon] \geq 1 - \delta$$

In the case of a density vector Q, we use the L_1 distance between vectors. The algorithm A approximates Q if $Prob_\Omega[\| Q - A(s) \|_1 \leq \varepsilon] \geq 1 - \delta$. The randomized algorithm A takes samples $t \in I$ from the stream with different distributions, introduced in the next section.

In the case of a community detection, the algorithm A gives a 1 (there is a large cluster) or 0 (there is no large cluster). In this case, the algorithm A accepts or rejects and we require that:

$$Prob_\Omega[A(s) \text{ is correct}] \geq 1 - \delta$$

We may also want to approximate a large community $S \subseteq V$ in a graph $G = (V, E)$ with a set $C \subseteq V$ which intersects S. The function $F : \Sigma^* \to 2^V$ takes a stream s of edges as input and $F(s) \subseteq V$. The algorithm A δ-approximates the function F if for all s,

$$Prob_\Omega[A(s) \cap F(s) \neq \emptyset] \geq 1 - \delta$$

The randomized algorithm A takes sample edges from the stream s with a uniform distribution and outputs a large subset $A(s) = C$ of the nodes. All our algorithms require a space of $k . \log | V |$, i.e. $O(k . \log n)$ for a graph with n nodes.

Reservoir Sampling. A classical technique, introduced in [16] is to sample each new tuple (edge) of a stream s with some probability p and to keep it in a set S called the *Reservoir* which holds k tuples. In the case of tuples t of a Datawarehouse with a measure $t.M$, we keep them with a probability proportional to their measures.

Let $s = t_1, t_2, ...t_n$ be the stream of tuples t with the measure $t.M$, and let $T_n = \sum_{i=1,...n} t_i.M$ and let $\widehat{S_n}$ be the Reservoir at stage n. We write $\widehat{S}$ to denote that S is a random variable.

k-Reservoir sampling: A(s)

- Initialize $S_k = \{t_1, t_2, ...t_k\}$,
- For $j = k + 1, ...n$, select t_j with probability $(k * t_j.M)/T_j$. If it is selected replace a random element of the Reservoir (with probability $1/k$) by t_j.

The key property is that each tuple t_i is taken proportionally to its measure. It is a classical simple argument which we recall.

Lemma 1. *Let S_n be the Reservoir at stage n. Then for all $n > k$ and $1 \leq i \leq n$:*

$$Prob[t_i \in S_n] = k.t_i.M/T_n$$

Proof. Let us prove by induction on n. The probability at stage $n + 1$ that t_i is in the Reservoir $Prob[t_i \in S_{n+1}]$ is composed of two events: either the tuple t_{n+1} does not enter the Reservoir, with probability $(1 - k.t_{n+1}/T_{n+1})$ or the tuple t_{n+1} enters the Reservoir with probability $k.t_{n+1}/T_{n+1}$ and the tuple t_i is maintained with probability $(k - 1)/k$. Hence:

$$Prob[t_i \in S_{n+1}] = k.t_i.M/T_n((1 - k.t_{n+1}/T_{n+1}) + k.t_{n+1}/T_{n+1} .(k - 1)/k)$$
$$Prob[t_i \in S_{n+1}] = k.t_i.M/T_n(1 - t_{n+1}/T_{n+1}) = k.t_i.M/T_{n+1}$$

In the case of edges, the measure is always 1 and all the edges are uniform.

Dynamic Reservoir Sampling. It is natural to analyze the most recent tuples in a stream. A sliding window [13] of length τ only considers at time t the tuples between time $t-\tau$ and time τ. New tuples arrive and old tuples leave the window. Alternative methods to generalize the Reservoir sampling are presented in [5]. A priority sampling (or MinHash) assigns a random value in the $[0, 1]$ interval to each tuple and selects the tuple with the minimum value. Each tuple is then selected with the uniform distribution. We assume that k tuples can be efficiently chosen in a sliding window.

In [1], a biased Reservoir sampling is introduced to penalize older tuples, another alternative to the sliding window.

2.4   Other Approaches

There are many other approaches to detect large clusters in streams of graphs edges, which first need a model of dynamic graphs. The *dynamic graphs algorithms* community such as [7] studies the compromise between update and query time in the worst case. The *graph streaming approach* [13] emphasizes the space complexity in the worst case and in particular for the window model. The *network sampling approach* such as [4] does consider the uniform sampling on the edges but there is no analysis for the detection of clusters in dynamic graphs. The detection of a *planted clique* is a classical problem [11], hard when the clique size is for example $O(\sqrt{(n)}/2)$ in the worst case. The *graph mining community* [2,3] studies the detection of clusters with different models.

In our approach, we only consider classes of graphs which follow a power law degree distribution and study approximate algorithms for the detection of large α-cliques in the window model.

3   Streaming Datawarehouse and Approximate OLAP

Assume a stream of tuples where the measure M is always bounded. Two important methods can be used to sample a Datawarehouse stream I:

- Uniform sampling: we select $\widehat{I}$, made of k distinct samples of I, with a uniform Reservoir sampling on the m tuples,
- Weighted sampling: we select $\widehat{I}$ made of k distinct samples of I, with a k-*weighted Reservoir sampling* on the m tuples. The measure of the samples is removed, i.e. reset to 1.

We concentrate on a k-weighted Reservoir. Let $\widehat{Q_C}$ be the density of Q_C on $\widehat{I}$ as represented in Fig. 2, with the weighted sampling, i.e. $\widehat{Q}_{C=c}$ be the density of Q on the value c of the dimension C, i.e. the number of samples such that $C = c$ divided by k. We simply interpret the samples with a measure of 1, i.e. computes $\widehat{Q}_C$. Let $\mid C \mid$ be the number of possible values c of the attribute C.

In order to show that $\widehat{Q}_C$ is an (ε, δ)-approximation of Q_C, we look at each component $Q_{C=c}$. We show that $\mathbb{E}(\widehat{Q}_{C=c})$ the expected value of $\widehat{Q}_{C=c}$ is $Q_{C=c}$. We then apply a Chernoff bound and a union bound.

Lemma 2. *Q_C, i.e. the density of Q on the dimension C can be (ε, δ)-approximated by $\widehat{Q}_C$ if $k \geq \frac{1}{2}.(\frac{|C|}{\varepsilon})^2. \log \frac{1}{\delta}$.*

Proof. Let us evaluate $\mathbb{E}(\widehat{Q}_{C=c})$, the expectation of the density of the samples. It is the expected number of samples with $C = c$ divided by k the total number of samples. The expected number of samples is $\sum_{t:t.C=c} \frac{k.t.M}{T}$ as each t such that $C = c$ is taken with probability $\frac{k.t.M}{T}$ by the weighted Reservoir for any total weight T. Therefore:

$$\mathbb{E}(\widehat{Q}_{C=c}) = \frac{\sum_{t:t.C=c} \frac{k.t.M}{T}}{k} = \frac{\sum_{t:t.C=c} t.M}{T} = Q_{C=c}$$

i.e. the expectation of the density $\widehat{Q}_{C=c}$ is precisely $Q_{C=c}$. As the tuples of the Reservoir are taken *independently* and as the densities are less than 1, we can apply a Chernoff-Hoeffding bound [12]:

$$Prob[|\, Q_{C=c} - I\!\!E(\widehat{Q}_{C=c})\,| \geq t] \leq e^{-2t^2.k}$$

In this form, t is the error and $1 - \delta = 1 - e^{-2t^2.k}$ is the confidence. We set $t = \frac{\epsilon}{|C|}$, and $\delta = e^{-2t^2.k}$. We apply the previous inequality for all $c \in C$. With a union bound, we conclude that if $k > \frac{1}{2}.(\frac{|C|}{\epsilon})^2. \log \frac{1}{\delta}$ then:

$$Prob[|\, Q_C - I\!\!E(\widehat{Q_C})\,| \leq \varepsilon] \geq 1 - \delta$$

This result generalizes to arbitrary dimensions but is of limited use in practice. If the OLAP query has a selection σ, the result will not hold because the samples will not follow the right distribution after the selection. If we sample the stream after we apply the selection, it will hold again. Hence we need to combine sampling and composition operations in a non trivial way.

In particular, if we combine two Datawarehouses with a new schema, it is difficult to correctly sample the two streams. In the case of two graphs, i.e. a simpler case, we propose a solution in the next section.

4   Streaming Graphs

Let $e_1, e_2, ...e_i...$ be a stream of edges where each $e_i = (u, v)$. It defines a graph $G = (V, E)$ where V is the set of nodes and $E \subseteq V^2$ is the set of edges: we allow self-loops and multi edges and assume that the graph is symmetric. In the *window model* we isolate the most recent edges at some discrete $t_1, t_2, ...$ We fix the length of the window τ, hence $t_1 = \tau$ and each $t_i = \tau + \lambda.(i-1)$ for $i > 1$ and $\lambda < \tau$ determines a window of length τ and a graph G_i defined by the edges in the window or time interval $[t_i - \tau, t_i]$. We therefore generate a sequence of graphs $G_1, G_2, ...$ at times $t_1, t_2, ...$ The graphs G_{i+1} and G_i share many edges: old edges of G_i are removed and new edges are added to G_{i+1}. Social graphs have a specific structure, a specific degree distribution (power law), a small diameter and some dense clusters. The dynamic random graphs introduced in this section satisfy these conditions and in this model, we can prove that a simple Community detection algorithm detects large enough Communities with high probability.

4.1   Random Graphs

The most classical model of random graphs is the Erdös-Renyi $G(n, p)$ model (see [9]) where V is a set of n nodes and each edge $e = (i, j)$ is chosen independently with probability p.

In the Preferential Attachment model, $PA(m)$, (see [6]), the random graph $\widehat{G}_n$ with n nodes is built dynamically: given $\widehat{G}_n$ at stage n, we build $\widehat{G}_{n+1}$ by adding a new node and m edges connecting the new node with a random node

j following the degree distribution in $\widehat{G}_n$. The resulting graphs have a degree distribution which follows a power law, such as the Zipfian law where:

$$Prob[d(i) = j] = \frac{c}{j^2}$$

In this case the maximum degree is $\sqrt{c.n}$ and the average degree is $c. \log n$, hence $m = O(c.n. \log n)$.

The Configuration Model [15] fixes an arbitrary degree distribution $\mathcal{D} = [D(1), D(2), ...D(k)]$ where $D(i)$ is the number of nodes of degree i. It generates a random graph among all the graphs with $\sum_i D(i)$ nodes and $\sum_i i * D(i)/2$ edges. For example if $\delta = [4, 3, 2]^1$, i.e. approximately a power law, we search for a graph with 9 nodes and 8 edges. Specifically 4 nodes of degree 1, 3 nodes of degree 2 and 2 nodes of degree 3, as in Fig. 3(a). Alternatively, we may represent $\mathcal{D}$ as a distribution, i.e. $\mathcal{D} = [\frac{4}{9}, \frac{1}{3}, \frac{2}{9}]$.

The configuration model generates graphs with the degree distribution $\mathcal{D}$ when $\sum_i i * D(i)$ is even and $\mathcal{D}$ satisfies the Erdös-Gallai condition [10]. Enumerate the half-edges (stubs) of the nodes according to their degrees, and select a random symetric matching π between the half-edges such that $\pi(i) \neq i$. The graph may have self-loops and multiple edges. If $\mathcal{D}$ follows a Zipfian power law, then the maximum degree is $O(\sqrt{m})$ if the graph has m edges. If π is selected with the uniform distribution, we say that the graph G is chosen by the *uniform selection*. A typical graph generated by the uniform selection is not likely to have a large cluster.

Consider the *S-concentrated selection*: fix some a subset S among the nodes of high degree. With probability α (say 80%), match the stubs in S uniformly in S, and with probability $1-\alpha$ (say 20%), match the stubs in S uniformly in $V - S$. Match the remaining stubs uniformly. This process concentrates edges in S and will create an α-cluster with high probability, assuming the degree distribution is a power law.

A classical study in random graphs is to find sufficient conditions so that the random graph has a *giant component*, i.e. of size $O(n)$ for a graph of size n. In the Erdös-Renyi model $G(n, p)$, it requires that $p > 1/n$, and in the Configuration Model $\mathbb{E}[\mathcal{D}^2] - 2\mathbb{E}[\mathcal{D}] > 0$ which is realized for the Zipfian distribution. There is a phase transition for both models. The size of the connected components in a graph specified by a degree sequence is studied in [8].

We can extend the two selection mechanisms to dynamic random graphs. Remove random $q \geq 2$ edges from G and reapply one of the selections. The *Uniform Dynamics* is obtained when we always choose the uniform selection, and the *S-concentrated Dynamics* is obtained when we always choose the S-concentrated selection. In the *Step Dynamics* we start with the uniform selection, then choose the S-concentrated selection during a time Δ and switch back to the uniform selection.

---

[1] Alternatively, one may give a sequence of integers, the degrees of the various nodes in increasing order, i.e. $[1, 1, 1, 1, 2, 2, 2, 3, 3]$, a sequence of length 9 for the distribution $\delta = [4, 3, 2]$.

4.2   Random Graphs with p Communities

The uniform selection will only generate very small communities. In contrast, the S-concentrated selection generates a large community, of size $|S|$ with high probability. In order to generate several communities, there are at least two models. The first model is to generalize the S-concentrated selection, to an S_1, S_2-concentrated selection, for disjoint sets S_1, S_2 of vertices. We first march the stubs of S_1 in S_1 with probability 80%, then the free stubs of S_2 in S_2 with probability 80% and the remaining stubs uniformly. It generalizes to p communities.

Another method is to generate $\widehat{G}_1$ and $\widehat{G}_2$ of the same size following the S-concentrated selection when $\mathcal{D}$ follows a power law. Let

$$\widehat{G} = \widehat{G}_1 \mid \widehat{G}_2$$

be the new graph $\widehat{G}$ which is the union of $\widehat{G}_1$ and $\widehat{G}_2$ with a few random edges connecting the nodes of low degree. This construction exhibits two communities S_1 and S_2 and its degree distribution is close to $\mathcal{D}$. It generalizes to p communities of different sizes, as in Fig. 3.

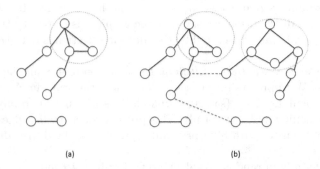

(a)   (b)

Fig. 3. Concentrated random graph for D with one community in (a). Random graph close to D with 2 communities in (b) where $\delta = [4, 3, 2]$ (or $[\frac{4}{9}, \frac{1}{3}, \frac{2}{9}]$ as a distribution).

4.3   Community Detection

A graph has a community structure if the nodes can be grouped into p disjoint dense subgraphs. Given a graph $G = (V, E)$, we want to partition V into $p + 1$ components, such that $V = V_1 \oplus V_2.... \oplus V_p \oplus V_{p+1}$ where each V_i for $1 \leq i \leq p$ is dense, i.e. $|E_i| \geq \alpha_i.|V_i|^2$ for some constants α_i, and E_i is the set of edges connecting nodes of V_i. In addition, the V_i do not have small cuts. The set V_{p+1} groups nodes which are not parts of the communities.

In the simplest case of 1 component, $V = V_1 \oplus V_2$ such that V_1 is dense and V_2 is the set of unclassified nodes. We first decide whether or not there is community and simply check if the largest connected component is of size greater than a constant h.

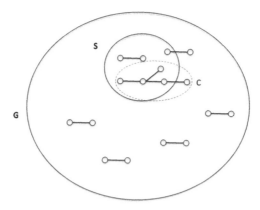

Fig. 4. Concentrated random graph G with a community S and 10 random edges from the Reservoir defining $\widehat{G}$ with the large connected component C with 4 edges.

Algorithm $A_1(k, h)$: Community detection Algorithm for a stream s of m edges.

- Maintain a k-Reservoir,
- Let C be the largest connected component of the k-Reservoir. If $|C| \geq h$ then Accept, else Reject

In a typical example, $m = 2.10^4$, $k = 400$, $h = 10$. Each large C_i will contain nodes of high degrees, and we will interpret C_i as a community. Figure 5 is an example of the connected components of the Reservoir. We now prove that this simple detection algorithm detects with high probability if we used the uniform or the S-concentrated selection, when S is large enough, with high probability (Fig. 4).

Lemma 3. *For the S-concentrated selection and m large enough, if $|S| \geq m/\alpha.k$, then there is a β such that $Prob_\Omega[|C| > \beta.|S|] \geq 1 - \delta$.*

Proof. If S is a clique, each edge is selected with constant probability k/m, so we are in the case of the Erdös-Renyi model $G(n, p)$ where $n = |S|$ and $p = k/m$. We know that the phase transition occurs at $p = 1/n$, i.e. there is a giant component if $p > 1/n$, i.e. $|S| \geq m/k$.

If S is almost a clique, i.e. an α-cluster, then the phase transition occurs at $p = 1/\alpha.|S|$. Hence if $p > 1/\alpha.|S|$, there is a giant component of size larger than a constant times $|S|$, with high probability $1 - \delta$. As the probability of the edges $p = k/m$, it occurs if $|S| \geq m/\alpha.k$. In this case, $|C|$ is $O(|S|)$. Hence for m large enough, there is constant β such that $Prob_\Omega[|C| > \beta.|S|] \geq 1 - \delta$.

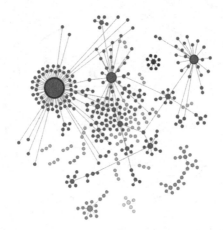

Fig. 5. Connected components of the k-Reservoir.

Notice that k must be large, i.e. $O(\sqrt{n}.\log n)$, hence $k/m = O(1/\sqrt{n})$ and $|S| \geq m/\alpha.k = O(\sqrt{n}/\alpha)$. For the uniform selection, there is an h such that $Prob_\Omega[|C| < h] \geq 1 - \delta$, using the result of [14]. We can now prove the main result:

Theorem 1. *Let G be a graph defined by a stream of m edges following a power law $\mathcal{D}$ and $k = O(\sqrt{n}.\log n)$. For the S-concentrated selection, if $|S| \geq m/\alpha.k$ and m large enough, then $Prob_\Omega[A_1$ Accepts$] \geq 1-\delta$ For the Uniform Dynamics $Prob_\Omega[A_1$ Rejects$] \geq 1 - \delta$.*

Proof. If $|S| \geq m/\alpha.k$ for the concentrated Dynamics, Lemma 3 states that $|C| > h$ with high probability, hence $Prob_\Omega[$Algorithm 1 Accepts$] \geq 1 - \delta$. For the Uniform Dynamics (equivalent to $|S| = 0$), there are no giant components in the Reservoir, hence $Prob_\Omega[$Algorithm 1 Rejects$] \geq 1 - \delta$.

4.4 Dynamic Community Detection

We extend the community detection algorithm with a dynamic Reservoir adapted to a sliding window. In the *window model* we isolate the most recent edges at some discrete $\tau_1, \tau_2, ...$ We fix the length of the window τ, hence $\tau_1 = \tau$ and each $\tau_i = \tau + \lambda.(i-1)$ for $i > 1$ and $\lambda < \tau$ determines a window of length τ and a graph G_i defined by the edges in the window or time interval $[\tau_i - \tau, \tau_i]$. We therefore generate a sequence of graphs $G_1, G_2, ...$ at times $\tau_1, \tau_2, ...$ The graphs G_{i+1} and G_i share many edges: old edges of G_i are removed and new edges are added to G_{i+1}.

We generalize the uniform and S-concentrated selection to *Dynamics*, where the graphs evolve in time. At time τ_i remove q random edges $(q > 2)$, i.e.

2.q stubs, and reapply the uniform selection or the S-concentrated selection. A *Dynamics* is a function which decides which of the two selection rules to apply at times $\tau_1, \tau_2, ...$ An example is the **Step Dynamics:** apply the Uniform Dynamics first, then switch to the S-dynamics for a time period Δ, and switch back to the Uniform Dynamics. The Uniform Dynamics is obtained when we always use the uniform selection and the S concentrated Dynamics when we only use the S-concentrated selection. In our setting, the Dynamics depends on some external information, which we try to approximately recover.

We store the large connected components of the Reservoir in a data structure described in the next section.

Algorithm $A_2(k, h, \tau, \lambda)$: Dynamic Community Algorithm for a stream s of edges.

- Maintain a dynamic k-Reservoir window of length τ,
- Store the large (of size greater than h) connected components $C_{i,1}, ... C_{i,l}$ of the k-Reservoir window at time τ_i for a time interval of length τ.

Suppose we are only interested in detecting a community. We generalize the static algorithm $A_1(k, h)$ into $A_3(k, h, \tau, \lambda)$:

Algorithm $A_3(k, h, \tau, \lambda)$: Dynamic Community Detection Algorithm.

- Maintain a dynamic k-Reservoir window of length τ,
- If there is a time τ_i such that the largest connected component $C_{i,j}$ is of size greater than h, Accept else Reject.

We can still distinguish between the Uniform and the S-concentrated Dynamics, if S is large enough. Let $m(t)$ be the number of edges in the window at time t.

Theorem 2. *Let $G(t)$ be a graph defined by a stream of $m(t)$ edges following a power law $\mathcal{D}$. For the S-concentrated Dynamics, if $|S| \geq m(t)/\alpha.k$ and $m(t)$ large enough, then $Prob_\Omega[A_3$ Accepts$] \geq 1 - \delta^{t/\tau}$ For the Uniform Dynamics $Prob_\Omega[A_3$ Rejects$] \geq (1 - \delta)^{t/\lambda}$.*

Proof. For each window, we can apply Theorem 1 and there are t/τ independent windows. If $|S| \geq m/\alpha.k$ for the concentrated Dynamics, the error probability is smaller than the error made for t/τ independent windows, which is $\delta^{t/\tau}$. Hence $Prob_\Omega[$Algorithm 3 Accepts$] \geq 1 - \delta^{t/\lambda}$. For the Uniform Dynamics (equivalent to $|S| = 0$), the algorithm needs to be correct at each t/λ step. Hence $Prob_\Omega[$Algorithm 3 Rejects$] \geq (1 - \delta)^{t/\lambda}$.

The probability to accept for the S concentrated Dynamics is amplified whereas the probability to reject for the Uniform Dynamics decreases. One single error generates a global error. In the implementation, we follow for 24 h a sliding window of length $\tau = 1$ h and $\lambda = 30$ mins. Figure 8 shows the dynamic evolution of the sizes of the communities.

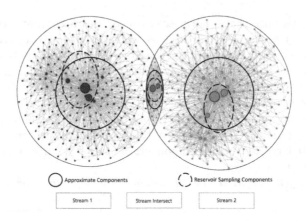

Fig. 6. Common communities between two graphs

5   Integration from Multiple Sources

Given two streams of edges defining two graphs $G_i = (V_i, E_i)$ for $i = 1, 2$, what is the integration and the correlation of these two structures? A first approach to their correlation would be to consider the Jaccard similarity[2] $J(V_1, V_2)$ on the domains of the two graphs. It has several drawbacks: it is independent of the structures of the graphs, it is very sensitive to noise, nodes connected with one or few edges and it is not well adapted when the sizes are very different.

We propose instead the following approach: we first estimate the dense components (clusters or communities) and apply the Jaccard similarity only to these dense components. It exploits the structure of the graphs, is insensitive to noise and adapts well when the graphs have different sizes. Let $C_i = \bigcup_j C_{i,j}$ be the set of clusters of the graph G_i, for $i = 1$ or 2. The *Content Correlation of two graphs* $\rho = J(C_1, C_2)$. This definition is $\rho(\tau_1)$ for the first window or for two static graphs. For two dynamic streams $G_1(t)$ and $G_2(t)$ which share a time scale, we generalize C_i into $C_i(t) = \bigcup_j C_{i,j}(t)$. *The Content Correlation of two graph streams* $G_1(t)$ *and* $G_2(t)$ *is*

$$\rho(t) = J(C_1(t), C_2(t))$$

In this definition, a Correlation gives the same weight to all the components, independently of the time. We can refine it, as an *amortized Correlation* $\rho_a(t)$. Let u be a node in $C_1(t) \cap C_2(t)$. It can appear in several components of $C_1(t)$ and $C_2(t)$. Let $\tau_{u,1}$ (resp. $\tau_{u,2}$) be the most recent time u appears in $C_1(t)$ (resp. in $C_2(t)$). Suppose $\tau_{u,1} > \tau_{u,2}$, and $\Delta(u) = \tau_{u,1} - \tau_{u,2}$. Let the weight $w(u)$ of u at time t be $w(u) = (1 - \frac{\Delta(u)}{t}) \cdot (\frac{\tau_{u,1}}{t})$. The weight is 1 when u occurs in the

---

[2] The Jaccard similarity or Index between two sets A and B is $J(A, B) = |A \cap B| / |A \cup B|$. The Jaccard distance is $1 - J(A, B)$.

most recent components of $G_1(t)$ and $G_2(t)$. It decreases to 0 if $\Delta(u)$ is large or if the most recent time of occurrence $\tau_{u,1}$ is far from t. Let

$$\rho_a(t) = \frac{\sum_{u \in C_1(t) \cap C_2(t)} w(u)}{|C_1(t) \cap C_2(t)|}$$

5.1 Integration of the Streams

At some discrete times $\tau_1, \tau_2, \ldots,$ we store the large connected components of the Reservoirs R_t of each stream. There could be no component stored. We use a NoSQL database, with 4 (Key, Values) tables where the key is always a tag (@x or #y) and the Values store the clusters nodes.

A stream is identified by a tag (or a set of tags) and a cluster is also identified by a tag, its node of highest degree. Hence some tags are possibly nodes, clusters and streams. The (Key, Values) tables are:

– *Stream(tag, list(cluster, timestamp))* provides the clusters of a stream,
– *Cluster(tag, list(stream, timestamp, list(high-degree nodes), list(nodes, degree)))* provides for a given cluster, the list of high-degree nodes and the list of nodes with their degree,
– *Nodes(tag, list(stream, cluster, timestamp))* provides for each node the list of streams, clusters and timestamps where it appears,
– *Correlation((tag1,tag2), list(value,timestamp))* the table which provides for each pair of streams *(tag1,tag2)* the different correlation values $\rho(t)$.

This data stucture is independent of the number of streams. It stores very limited information and is the *integration* of several streams.

5.2 Approximate Correlation Between Two Streams

Consider the following online algorithm to compute $\rho(t + \lambda)$, given $\rho(t)$ and the sets $C_i(t)$ for $i = 1, 2$.

Online Algorithm $A_4(h)$ for ρ. Assume $\rho(t) = I/U$ where by definition $I = |C_1(t) \cap C_2(t)|$ and $U = |C_1(t) \cup C_2(t)|$. At time $t + \lambda$:

– Enumerate the new large connected components (of size greater than h)
– Compute the increase δ_i in size of $C_i(t + \lambda)$ for $i = 1, 2$ from $C_i(t)$,
– Compute the increase δ' in size of $C_1(t + \lambda) \cap C_2(t + \lambda)$.
– Then $\rho(t + \lambda) = \rho(t) + \frac{U.\delta' - I.(\delta_1 + \delta_2)}{U.(U + \delta_1 + \delta_2)}$.

A simple computation shows that $\rho(t + \lambda) = \frac{I + \delta'}{U + \delta_1 + \delta_2}$, i.e. the correct definition. There are several direct applications.

Suppose we have two streams $G_1(t)$ and $G_2(t)$ which share the same clock. Suppose that $G_1(t)$ is a step strategy Δ_1 on a cluster S_1 and $G_2(t)$ is a step strategy Δ_2 on a cluster S_2. Let $\rho^* = J(S_1, S_2)$. How good is the estimation of their correlation? Let $C_i(t) = \bigcup_j C_{i,j}$ be the set of large clusters $C_{i,j}$ of the graph G_i, for $i = 1$ or 2 at time t.

Theorem 3. *For m large enough, $G_1(t)$ and $G_2(t)$ two step strategies before time t on two clusters such that $|S_i| \geq m/\alpha.k$ for $i = 1, 2$. Then $Prob_{\Omega_t}[\rho(t) - \rho^* \leq \varepsilon] \geq 1 - \delta$.*

Proof. After the first observed step, for example on S_1, Lemma 3 indicates that the domain of the largest connected component V_{S_1} is already some approximation of S_1. After Δ_1/τ independent trials, $V_1 = \bigcup_i V_{S_{1,i}}$ will be a good approximation of S_1. Similarly for S_2 and therefore $\rho(t) = J(V_1, V_2)$ will (ε, δ) approximate ρ^*.

6   Experiments

Given a set of tags such as #CNN or #Bitcoin, Twitter provides a stream of tweets represented as Json trees whose content contains these tags. The *Twitter Graph* of the stream was introduced in Sect. 2.2. A stream of tweets is then transformed into a stream of edges $e_1, ...e_m, ...$

We captured 4 twitter streams[3] on the tags #CNN, #FoxNews, #Bitcoin, and #Ripple during 24 h with a window size of $\tau = 1$ h, a time interval $\lambda = 30$ mins, a Reservoir size $k = 400$ and a threshold value $h = 10$. Figure 7 indicates the number of edges seen in a window, approximately $m = 20.10^3$ per stream, on 48 points. We have 24 independent windows, read approximately 48.10^4 edges on each stream, and globally approximately 2.10^6 edges. On the average we save 100 nodes, i.e. $4.48.100 \simeq 10^4$ nodes, i.e. a compression of 200. For $\alpha = 0.8$, the minimum size associated with the concentrated model is $m/\alpha.k \simeq 60$.

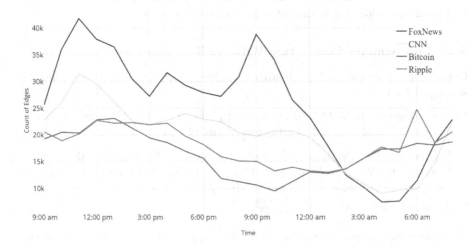

Fig. 7. Number of edges in a window of 1 h, for 4 streams, during 24 h (saved every 30 mins)

---

[3] Using a program available on https://github.com/twitterUP2/stream which takes some tags, a Reservoir size k, a window size τ, an interval value λ and saves the large connected components of the Twitter stream.

6.1   Stability of the Components

As we observe the dynamics of the communities, there is some instability: some components appear, disappear and may reappear later, as Fig. 8 shows.

The sizes of the large components are quite stable in time as Fig. 9 shows.

It is best observed with the following experiment: assume two independent Reservoirs of size $k' = k/2$ as in Fig. 10. More complete results can be found at

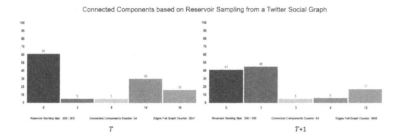

Fig. 8. Sizes of the connected components online

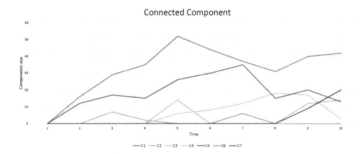

Fig. 9. Evolution of the sizes of the connected components

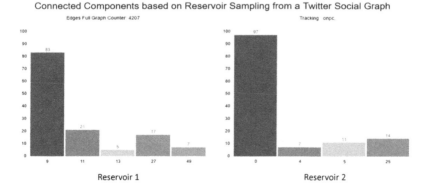

Fig. 10. Sizes of the connected components with 2 independent Reservoirs

http://www.up2.fr/twitter. The last two communities of the Reservoir 1 with 5 communities merge to correspond to the 4 communities in Reservoir 2.

Small communities $C_{i,j}$ are most likely trees, hence unstable as the removal of 1 edge splits the component and the sizes drop below the threshold h. Larger components are graphs which are therefore more stable.

6.2   Correlations

Figure 11 gives the three mains correlations $\rho(t)$ out of the possible 6 for the 4 streams. Figure 12 gives the averaged correlation

$$\rho'(t) = (\rho(t-1) + \rho(t) + \rho(t+1))/3$$

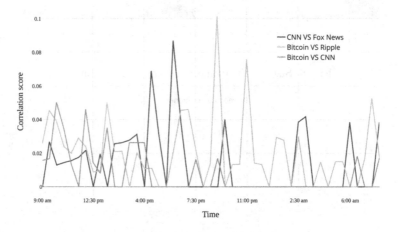

Fig. 11. Evolution of the stream correlation

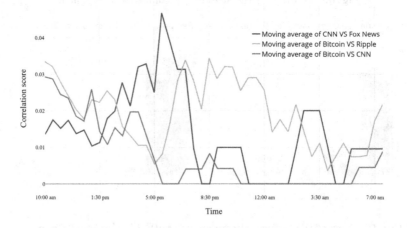

Fig. 12. Evolution of the averaged stream correlation

as the direct correlation is highly discontinuous. As expected, the averaged version is smoother. The maximum value is 1% for the correlation and 0.5% for the averaged version. It is always small as witnessed by the correlation matrix of Fig. 13. The spectrum of the Reservoirs, i.e. the sizes of the large connected components is another interesting indicator. For the #Bitcoin stream, there is a unique very large component.

$$
\begin{array}{cccc}
 & \textit{CNN} & \textit{Fox News} & \text{Bitcoin} & \text{Ripple} \\
\textit{CNN} & 1 & 0.06 & 0.009 & 0 \\
\textit{Fox News} & 0.06 & 1 & 0 & 0 \\
\text{Bitcoin} & 0.009 & 0 & 1 & 0.08 \\
\text{Ripple} & 0 & 0 & 0.08 & 1
\end{array}
$$

Fig. 13. Matrix correlation between 4 streams

7   Conclusion

We presented approximation algorithms for streams of tuples of a Datawarehouse and for streams of edges of a Social graph. Our algorithms store k-edges in a Reservoir from m edges and each edge is selected with probability k/m. They use $O(k.\log n)$ space whereas the entire graph may need $O(n^2.\log n)$ space.

If there is a large enough community, it will appear as a large connected component in the Reservoir, using the Erdös-Renyi model $G(n, p)$ phase transition. If there is no large community the largest connected component will be small. This basic community detection algorithm is an approximate algorithm on graphs which follow a power law degree distribution. It generalizes to dynamic graphs.

The Content Correlation of two streams of graph edges is the fraction of the nodes in common communities. It is the basis for the Integration of two streams of edges which only stores the large communities of streams. The Correlation can then be computed online.

We illustrated this approach with Twitter streams associated with 4 keywords and estimated their Correlations with this method.

References

1. Aggarwal, C.C.: On biased reservoir sampling in the presence of stream evolution. In: Proceedings of the 32nd International Conference on Very Large Data Bases, VLDB 2006, pp. 607–618 (2006)
2. Aggarwal, C.C.: An introduction to cluster analysis. In: Data Clustering: Algorithms and Applications, pp. 1–28. CRC Press (2013)
3. Aggarwal, C.C., Zhao, Y., Yu, P.S.: On clustering graph streams. In: Proceedings of the SIAM International Conference on Data Mining, SDM 2010, 29 April–1 May 2010, Columbus, Ohio, USA, pp. 478–489 (2010)

4. Ahmed, N.K., Neville, J., Kompella, R.: Network sampling: from static to streaming graphs. ACM Trans. Knowl. Discov. Data **8**(2), 7:1–7:56 (2013)
5. Babcock, B., Datar, M., Motwani, R.: Sampling from a moving window over streaming data. In: Proceedings of the Thirteenth Annual ACM-SIAM Symposium on Discrete Algorithms, SODA 2002, pp. 633–634 (2002)
6. Barabasi, A., Albert, R.: The emergence of scaling in random networks. Science **286**, 509–512 (1999)
7. Bhattacharya, S., Henzinger, M., Nanongkai, D.: Fully dynamic approximate maximum matching and minimum vertex cover in $o(\log^3 n)$ worst case update time. CoRR and STOC 2016, arXiv:abs/1704.02844 (2017)
8. Chung, F., Lu, L.: Connected components in random graphs with given expected degree sequences. Ann. Comb. **6**, 125–145 (2002)
9. Erdös, P., Renyi, A.: On the evolution of random graphs. Publ. Math. Inst. Hung. Acad. Sci. **5**, 17–61 (1960)
10. Erdős, P., Gallai, T.: Gráfok előírt fokszámú pontokkal. Matematikai Lapok **11**, 264–274 (1960)
11. Feige, U., Krauthgamer, R.: Finding and certifying a large hidden clique in a semirandom graph. Random Struct. Algorithms **16**(2), 195–208 (2000)
12. Hoeffding, W.: Probability inequalities for sums of bounded random variables. J. Am. Stat. Assoc. **58**(2), 13–30 (1963)
13. McGregor, A.: Graph stream algorithms: a survey. SIGMOD Rec. **43**(1), 9–20 (2014)
14. Molloy, M., Reed, B.: The size of the giant component of a random graph with a given degree sequence. Comb. Probab. Comput. **7**(3), 295–305 (1998)
15. Newman, M.: Networks: An Introduction. Oxford University Press Inc., Oxford (2010)
16. Vitter, J.S.: Random sampling with a reservoir. ACM Trans. Math. Softw. **11**(1), 37–57 (1985)

Big Research Data Integration

Valentina Bartalesi$^{(\boxtimes)}$, Carlo Meghini, and Costantino Thanos

Istituto di Scienza e Tecnologie dell'Informazione "A. Faedo" – CNR, Pisa, Italy
{valentina.bartalesi,carlo.meghini,costantino.thanos}@isti.cnr.it

Abstract. The paper presents a vision about a new paradigm of data integration in the context of the scientific world, where data integration is instrumental in exploratory studies carried out by research teams. It briefly overviews the technological challenges to be faced in order to successfully carry out the traditional approach to data integration. Then, three important application scenarios are described in terms of their main characteristics that heavily influence the data integration process. The first application scenario is characterized by the need of large enterprises to combine information from a variety of heterogeneous data sets developed autonomously, managed and maintained independently from the others in the enterprises. The second application scenario is characterized by the need of many organizations to combine information from a large number of data sets dynamically created, distributed worldwide and available on the Web. The third application scenario is characterized by the need of scientists and researchers to connect each others research data as new insight is revealed by connections between diverse research data sets. The paper highlights the fact that the characteristics of the second and third application scenarios make unfeasible the traditional approach to data integration, i.e., the design of a global schema and mappings between the local schemata and the global schema. The focus of the paper is on the data integration problem in the context of the third application scenario. A new paradigm of data integration is proposed based on the emerging new empiricist scientific method, i.e., data driven research and the new data seeking paradigm, i.e., data exploration. Finally, a generic scientific application scenario is presented for the purpose of better illustrating the new data integration paradigm, and a concise list of actions that must be performed in order to successfully carry out the new paradigm of big research data integration is described.

Keywords: Research data integration · Big data · Semantic web

1 Introduction

Data Integration has the goal of enriching and completing the information available to the users by adding complementary information residing at diverse information sources [12,24]. It aims at providing a more comprehensive information basis in order to better satisfy user information needs. This is achieved by combining data residing at diverse data sets and creating a unified view of these

© Springer Nature Switzerland AG 2019
D. Kotzinos et al. (Eds.): ISIP 2018, CCIS 1040, pp. 23–37, 2019.
https://doi.org/10.1007/978-3-030-30284-9_2

datasets. This view provides a single access point to these distributed, hetero-geneous and autonomous data sets. Therefore, it frees the user from the neces-sity of interacting separately with each of these data sets. We distinguish two types of data integration [3]. The first type of data integration, structural data integration, refers to the ability to accommodate in a common data represen-tation model distributed data sets represented in different data representation models and formats. In essence, in this type of integration the goal is to aug-ment the dimensionality of an entity/object represented in different distributed data sets by collecting together all the attributes/features associated with this entity/object. The second type of data integration, semantic data integration [8], refers to the ability to combine distributed data sets on the basis of existing semantic relationships between them. In essence, in this type of integration the goal is to augment the relationality of an entity/object represented in a data set by linking it to entities/objects semantically closely related to it and represented in other distributed data sets.

The type of data integration very much depends on the characteristics of the data to be integrated. In the scientific domain, data can be referred to as raw or derivative. Raw data consist of original observations, such as those collected by satellite and beamed back to earth or generated by an instrument or sensor or collected by conducting an experiment. Derivative data are generated by processing activities. The raw data are frequently subject to subsequent stages of curation and analysis, depending on the research objectives. While the raw data may be the most complete form, derivative data may be more readily usable by others as processing usually makes data more usable, thus increasing their intelligibility. Structural data integration is, mainly, performed between data sets containing raw data, while semantic data integration is more appropriate for data sets containing derivative data; in this case, the semantic relationship between data sets is, usually, a correlation between them.

We have identified three main application scenarios where data integration is of paramount importance [1]. These three application scenarios well illustrate the evolution of the data integration concept both from the application and technological point of view. The first application scenario is characterized by the need of large enterprises to combine information from a variety of heterogeneous data sets developed autonomously, managed and maintained independently from the others in the enterprises. The second application scenario is characterized by the need of many organizations to combine information from a large num-ber of data sets dynamically created, distributed worldwide and available on the Web. The third application scenario is characterized by the need of scientists and researchers to connect each others research data as new insight is revealed by connections between diverse research data sets. In essence, we can say that data integration, in the first application scenario, is instrumental in effectively and efficiently managing large enterprises and in supporting the enterprise' planning activities. In the second application scenario, data integration is, mainly, instru-mental in activities like data mining, forecasting, statistical analysis, decision making, implementing strategy, etc., conducted by organizations whose business is based on the analysis and comparison of data stored in a large number of data collections distributed worldwide. In the third application scenario, data integration is, mainly, instrumental in exploratory studies carried out by research teams. For each one of the above three application scenarios, different technological challenges must be faced in order to develop integration systems that efficiently and effectively carry out the data integration process.

The paper is organized as follows: in Sect. 2, it overviews the technological challenges to be faced in order to successfully carry out the traditional data integration process. These challenges have been extensively discussed in the literature. In Sect. 3, the features of the three application scenarios that heavily influence the data integration process are described. In Sect. 4 a new paradigm of big research data integration is described. This is the main contribution of the paper. In 4 Subsections the enabling technologies are identified and briefly described. In Sect. 5 a generic scientific application scenario is presented for the purpose of better illustrating the new data integration paradigm. Finally, in Sect. 6, a concise list of actions that must be performed in order to successfully carry out the new paradigm of big research data integration is described.

2   Data Integration Technological Challenges

The traditional approach to data integration is a three-step process: data transformation, duplicate detection, and data fusion [19]. Data transformation is concerned with the transformation of the local data representations (local schemata) into a common representation, the global schema or mediated schema, which hides the structural aspects of the different local data collections. Two basic approaches have been proposed for this purpose [17]. The first approach, called Global-as-View (GAV), requires the global schema be expressed in terms of the local data schemata. In essence, this approach regards the local data schemata and generates a global schema that is complete and correct with respect to the local data schemata, and is also minimal and understandable. The second approach, called Local-as-View (LAV), requires the global schema be specified independently from the local data schemata, and the relationships between the global schema and the local data schemata are established by defining every local data schema as a view over the global schema.

The global schema can be materialized or virtual. In the first case, it is materialized in a persistent store, for example, in a data warehouse that consolidates data from multiple data sets. Extract-Transform-Load (ETL) tools [22] are used to extract, to transform, and load data from several data sets into a data warehouse. In the second case, the global schema is not materialized; it is virtual, that is, it just gives the illusion that the data sets have been integrated. The users pose queries against this virtual global schema; a mediator, i.e., a software module, translates these queries into queries against the single data sets and integrates the result of them. The second step, i.e., duplicate detection, is concerned with the identification of multiple, possibly, inconsistent representations of the same real-world entities. The third step, i.e., data fusion [19], is concerned with the fusion of the duplicate representations into a single representation and the resolution of the inconsistencies in the data.

Several technological challenges must be faced in order to efficiently and effectively carry out the data integration process. The first challenge to be faced regards the structural heterogeneity of the different data sets to be integrated. Integrating different data models and formats, i.e., structural data integration, requires the resolution of several types of conflicts. At the schema level, different data schemas may use (i) different data representations; (ii) different scales and measurement units; and (iii) different modelling choices, for example, an entity in one schema is represented as an attribute in another schema. At the data level, some contradictions may occur when different values exist for an attribute of the same entity. In addition, uncertainty may occur when a value of an attribute is missing in one data collection and is present in another data collection.

The second and more demanding challenge to be faced regards the semantic heterogeneity of the different data schemas to be integrated, i.e., semantic data integration. In general, the schemas of the different data sets do not provide explicit and precise semantics of the data to be integrated. The lack of precise data semantics can cause semantic ambiguities. For example, it may occur that two relations in two different data sets (assuming that both collections are modeled according to the relational data model) have the same name but heterogeneous semantics. This can induce in making erroneous design choices when the mediated schema is designed. In addition, the meanings of names and values may change over time. To mitigate the problem of semantic heterogeneity data must be endowed with appropriate metadata.

The third challenge to be faced regards the implementation of a mediating environment that provides a core set of intermediary services between the global schema and the local schemata. Such core set of services should include services that [23]:

- quickly and accurately find data that support specific user needs;
- map data structures, properties, and relationships from one data representation
- scheme to another one, equivalent from the semantic point of view;
- verify whether two strings/patterns match or whether semantically heterogeneous data match;
- optimize access strategies to provide small response time or low cost;
- resolve domain terminology and ontology differences; and
- prune data ranked low in quality or relevance.

In essence, the intermediary services must translate languages, data structures, logical representations, and concepts between global and local schemata. The effectiveness, efficiency, and computational complexity of the intermediary functions very much depend on the characteristics of the information models (expressiveness, semantic completeness, adequate modelling of data descriptive information (metadata), reasoning mechanisms, etc.) and languages adopted by the user. Ideally, they must provide a framework for semantics and reasoning. An important component of the mediating environment is ontologies [10]. Several domain-specific ontologies are being developed (gene ontology, sequence ontology, cell type ontology, biomedical ontology, CIDOC, etc.). Ontologies have been extensively used to support all the intermediary functions because they provide an explicit and machine-understandable conceptualization of a domain.

3   The Main Characteristics of the Three Application Scenarios

The first application scenario is characterized by:

- a relatively small number of data sets to be integrated;
- relatively static local schemas;
- fixed local schemas; and
- local schemas known in advance.

In such scenario, the design of the global schema as well as the mappings between the local schemas and the global schema are relatively easy tasks. Several data integration systems have been implemented which operate efficiently in this scenario [5, 24].

The second application scenario is characterized by:

- large number of data sets to be integrated;
- data sets containing huge volumes of data;
- data sets of widely differing data qualities;
- extremely heterogeneous local schemas;
- dynamically created local schemas (not fixed); and
- local schemas not known in advance.

In such an application environment, performing the data integration process is a very difficult task [9]. In fact:

- the large number of data sets to be integrated makes the alignment of their schemas very difficult;
- the extreme heterogeneity of data representation models adopted by the data sets to be integrated makes the design of a global schema very difficult;
- the dynamicity of the local schemas requires that the global schema undergoes continuous changes/extensions;
- the dynamicity of the local schemas makes difficult understanding the evolution of semantics and infeasible the capture of data changes timely;
- the widely differing qualities of the data to be integrated makes the alignment of the local schemas really hard; and
- the large volumes of the data to be integrated makes their warehousing very expensive.

In order to overcome the difficult problem of designing and maintaining a global schema in such a complex and dynamic context, it has been proposed to adopt an ontology-based approach to data management [7]. In this approach, the global schema is replaced by the conceptual model of the application domain. Such a model is formulated as an ontology expressed in a logic-based language. With this approach the integrated view is a semantically rich description of the relevant concepts in the domain of interest, as well as the relationships between such concepts. The users of an ontology-based data management system are enabled to query the data using the elements in the ontology as predicates. The ontology-based approach permits to overcome nicely the need for continuously reshaping of the global schema as an ontology can be easily extended. In fact, this approach supports an incremental process in representing the application domain. The domain ontology can be enriched with new concepts and relationships between them as new data sources or new elements in these sources are added. In essence, this approach supports the evolution of the ontology and the mappings between the ontology concepts and the data contained in the data sources supporting, thus, a "pay-as-you-go" data integration.

A conceptual difference between the above two data integration application scenarios regards the global perspective to be taken into account when designing global schemata. In the first application scenario, the designer of the global schema, by adopting the LAV approach, is enabled to take into due consideration a global perspective concerning the enterprise's activities. In the second application scenario, such an opportunity (i.e., the LAV approach) is not possible as the design of a global schema is practically unfeasible.

The third application scenario is situated within the scientific world. Indeed, we focused on the characteristics that are specific to data produced by research activities in the context of a new scientific framework characterized by; (i) the production of big data; and (ii) a science increasingly data intensive, multidisciplinary and e-science. In this world, some disciplines, for example, astronomy and high-energy physics, rely on a limited number of data repositories containing huge amounts of data. In that situation, researchers know where to find data of interest for their research activities. The problem is that the amount of data contained in these repositories outgrows the capabilities of query processing technology. In the case of overwhelming amounts of data, a new paradigm of query processing has been proposed: "data exploration" [14]. This new paradigm enables us to re-formulate the data integration problem as, mainly, a data interconnection problem. Exploration-based systems, instead of considering a huge data set in one go, incrementally and adaptively guide users towards the path that their queries and the result lead. These systems do not offer a correct and complete answer but rather a hint of how the data looks like and how to proceed further, i.e., what the next query should be. Data exploration, therefore, is a new approach in discovering connections and correlations between data in the big data era. Some other disciplines rely on large number of voluminous data sets with varying representation models, formats and semantics produced by many Labs and research groups distributed worldwide. Often, data of the same phenomenon come from many data sets. In such an application environment, discovering connections and correlations between data from autonomous distributed data sets is driving the need for data integration [6]. Integrating multiple data sets will enable science to advance more rapidly and in areas heretofore outside the realm of possibility. The third application scenario we consider addresses exactly the need for data integration of these scientific disciplines. Such an application scenario shares all the main characteristics of the second application scenario; in addition, it has some peculiar characteristics:

- the local schemata continuously evolve as new insights are gained in a scientific domain; for example, certain concepts can be invalidated in the light of new discoveries.
- data heterogeneity that is also created by the fact that researchers can conceptualize the same scientific problem in different ways due to the fact that, for example, belong to different "schools of thought".
- data heterogeneity that is intrinsic to some scientific disciplines. In fact, as reported in [16]: "in the environment of high-energy physics experiments (say, a particle detector), detector parts will be necessarily conceptualized differently depending on the kind of information system in which they are represented. For instance, in a CAD system that is used for designing the particle detector, parts will be spatial structures; in a construction management system, they will have to be represented as tree-like structures modeling compositions of parts and their sub-parts, and in simulation and experimental data taking, parts have to be aggregated by associated sensors (readout channels), with respect to which an experiment becomes a topological structure largely distinct from the one of the design drawing. We believe that such differences also lead to different views on the knowledge level, and certainly lead to different database schemata".
- data uncertainty is reported in different ways by different research communities.
- different data formats are adopted by different research communities.
- silos in modeling data sets; and
- different concepts of what to include in the metadata.

Therefore, the traditional approach to data integration based on the design of a global schema is unfeasible, also, in the third application scenario. In the next section we will outline a new paradigm of research data integration that is well suited to the way researchers are seeking scientific information.

4   Big Research Data Integration: A New Paradigm

TIn Sect. 3 we have sketched some characteristics of the Research Application Scenario that heavily influence the data integration process. In this Section, we extend the description of this Scenario with some other characteristics that are equally relevant for the data integration process. These characteristics are instrumental in the re-formulation of the data integration problem. The Big Research Data era is characterized by:

- huge volumes of data available in many fields of science;
- an increasingly production of new data types that augments the complexity of data sets;
- a worldwide distribution of data sets;
- data sets with high dynamism, uncertainty, exhaustivity, and relationality.

All these characteristics of big data have contributed to the emergence of new paradigms of seeking data and creating knowledge. We have, already, described in the previous Section the new paradigm of data seeking, that is, the data exploration. A new empiricist epistemological method [15] for creating new knowledge is also emerging. In the traditional scientific method, i.e., hypothesis driven research, the data are analyzed with a specific question in mind, that is, a hypothesis. In essence, this scientific method adopts a deductive reasoning for discovering new insights from the data. In the new empiricist method, i.e., data driven research, the data are analysed with no specific question in mind. In essence, huge volumes of data together with powerful analytic tools enable data to speak for themselves. Mining big data can reveal relationships and correlations that researchers did not even know to look for. In this method an inductive or abductive reasoning is adopted for discovering new insights from the data. These two new paradigms, i.e., data exploration and data driven research, heavily influence the data integration process too. The exploratory approach to data seeking suggests the possibility for researchers to start browsing in one data set and then navigating along links into related data sets, or to support data search engines to crawl the data space by following links between data sets The empiricist method entails the integration logic that must guide the creation of these links. In fact, in the hypothesis driven method a link between two data sets is established only when a semantic relationship between variables within these data sets, dictated by the hypothesis, holds. In the empiricist method a link between two data sets is established only when a correlation between variables within these data sets, is found. In essence, in the traditional approach the integration logic allows researchers to test a theory by analysing relevant data linked together on the basis of a deductive reasoning. In the empiricist approach, the integration logic enables researchers to discover new insights by analyzing data linked together on the basis of a inductive/abductive reasoning. Different kinds of logic (conventional logic, modal logic, causal logic, temporal logic, etc.) can be explored. An additional consideration that has an impact on the relationality/connectivity of a data set regards the properties of the data set relationships. They can be direct or indirect. A direct relationship between variables of two data sets can be recognized if these variables represent closely related concepts on the basis of the adopted logic reasoning. An indirect relationship between variables of two data sets can be established if they are directly related to a third mediating data set. Indirect relationships are based on direct relationships that enjoy the transitivity property, for example, the "causality" relationship is transitive. The indirect relationships increase the relationality/connectivity of the data sets and can enhance the data integration process and consequently the knowledge creation process.

The above two new paradigms make feasible the realization of significant advances in many scientific disciplines in the big data era as such advances can be driven by patterns in a data space. In fact, insights are arising from connections and correlations found between diverse types of data sets acquired from various modalities. Discovering semantic relationships between data sets enables new knowledge creation. The role of a data integration system is to make explicit hidden semantic relationships between data sets. Several types of semantic relationships can exist between object descriptions represented in database views. Examples of semantic relationships include: the inclusion relationship that is the standard subtype/supertype relationship; is-a and part-of relationships; member-collection relationship (association relationship); feature-event relationship; phase-activity relationship; place-area relationship; component-object relationship; antonyms/synonyms relationships, etc. Other types of semantic relationships can exist that are domain-specific. Making explicit hidden semantic relationships implies the creation of links between data sets. This process entails the creation of linked data spaces. Such linked data spaces can be implemented by exploiting linking technologies that allow to connect/link semantically related data sets. For example, data sets produced worldwide and related to the same phenomenon could be linked together creating, thus, linked data spaces in the form of thematic graphs. Researchers, interested in discovering correlations and semantic relationships between data sets contained in linked data spaces, should go through these thematic graphs. In essence, now the researchers have to explore a linked research data space by navigating through it. Based on all the above considerations the data integration paradigm problem can be re-formulated as follows:

Given a number of distributed heterogeneous and time varying data sets, link them on the basis of existing semantic and temporal relationships among them.

5   Enabling Technologies

IIn this section we briefly describe the main technologies that enable the implementation of the new paradigm of big research data integration as reformulated above. These technologies (Linked Open Data, Semantic Web technologies, vocabularies etc.) are tools which may be adequate to the realization of the paradigm or in some case may need to be extended the paradigm.

5.1   Data Abstraction/Database View

As, already, said research databases contain huge amounts of data. Usually, researchers are interested only in some parts of a database. These parts of a database (called sub datasets) are known as database views. A database view can be defined as a function [4] that, when applied to an instance (database) of a given database schema, produces a database in some other schema. In addition, the input and output schemata of this function could be represented in different data models. We think that each large database should be endowed with a number of (possibly overlapping) views. Therefore, linking database views, distributed worldwide, on the basis of semantic relationships existing between them will be instrumental in knowledge creation.

5.2 Data Citation

Citation systems [21] are of paramount importance for the discovery of knowledge in science as well as for the reproducibility of research outcomes. Indeed, being able to cite a research data set enables potential users to discover, access, understand and reproduce it. A citation is a collection of "snippets" of information (such as authorship, title, ownership, and date) that are specified by the administrator of the data set and that may be prescribed by some standards. In essence, a "snippet" of information constitutes the metadata that must be associated with each cited data set. A data citation capability should guarantee the uniquely identification of a data set. The unique identification is achieved by using persistent identifiers (PID) such as the Life Science Identifiers (LSID), the Digital Object Identifier (DOI), the Uniform Resource Name(URN), etc. A data citation capability should also guarantee that a data citation remains consistent over time, i.e., it has to show way to the original cited data set. Guaranteeing the persistence of a data citation is demanding when the data set to be cited evolves over time [20].

5.3 Semantic Web Technologies

Linked Open Data. The term Linked Open Data refers to a set of best practices for publishing structured data on the Web [13]. In particular, Linked Data provides (i) a unifying data model. Linked Data relies on Resource Description Framework RDF as a single, unifying model; (ii) a standardized data access mechanism. Linked Data commits itself to a specific pattern of using the HTTP protocol; (iii) hyperlink-based data discovery. By using URIs as global identifiers for entities, Linked Data allow hyperlinks to be set between entities in different data sources; and (iv) self-descriptive data. A grassroots effort, the Linked Open Data, is aiming to publish and interlink open license data sets from different data sources as Linked Data on the Web.

Resource Description Framework. Resource Description Framework (RDF) [18] is a language for representing information about resources that can be identified on the Web. It represents information as node-and-arc-labeled directed graphs. The data model is designed for the integrated representation of information that originates from multiple sources, is heterogeneously structured, and is represented using different schemata. RDF aims at being employed as a lingua franca, capable of moderating between other data models that are used on the Web. In RDF, a description of a resource is represented as a number of triples. The three parts of each triple are called its subject, predicate, and object.

Time in RDF. As most of the data in the Web are time varying, there is a need for representing temporal information in RDF. This will enable users to navigate in RDF graphs across time. Therefore, it will support queries that ask for past states of the data represented by the RDF graphs. Two main mechanisms for adding temporal information in RDF graphs have been proposed in the literature [11]. The first mechanism consists in time-stamping the RDF triples that are destined to change, i.e., adding a temporal element t that labels an RDF triple. The second mechanism consists in creating a new version of the RDF graph each time an RDF triple is changed.

Named RDF Graphs. The Named Graphs data model has been introduced in order to allow a more efficient representation of metadata information about RDF data and a globally unique identification of RDF data. It is a simple variation of the RDF data model. The main idea of the model is the introduction of a naming mechanism that allows RDF triples to talk about RDF graphs. A named graph is an entity that consists of an RDF graph and a name in the form of an URI reference.

6   A Generic Scientific Application Scenario

In this section a generic scientific application scenario is described and the data integration problem is refined. In this hypothetical application environment, there are n databases distributed worldwide containing heterogeneous data represented in different formats and managed by different data management systems. These data are produced by different research teams, each following its own practices and protocols:

DB1 DB2, . . . , DBn

Let's, also, suppose that several database views are defined on top of each database.

We further suppose that, by using special SW, a virtual layer consisting of RDF graphs is produced on top of these views. For example, we can produce an RDF view of a relational database schema [2]. A database view has an intention and an extension. The intention describes the semantics of the view, i.e., a data query embodying the schema of the database that is created when the query is applied to a database. The extension is the data subset defined by the schema of the view, that is, the data subset selected by the query each time the query is applied. For each view, there can be any number of extensions, each produced by applying the query at a different time. Each extension has the schema defined by the intension of the view. Views should be endowed with an identifier and domain-specific metadata. We suppose that the extensional definition of the database views varies over time while the intentional definition remains unchanged. Suppose that on top of each database a number of views Views (i) are defined. The notation View1(ti) indicates that the view named View1 was created at time ti. Therefore, the views

$$\text{View1(ti), View1(ti+x), View1(ti+x+1),\ldots, View1(ti+x+n)}$$

have all the same schema. The dynamic nature of the extensions of database views requires mechanisms that allow the tracing of all changes that occurred during the data subset life cycle. Such mechanisms should allow, when a database operation (insert/update/delete) modifies a database subset at a given time, the time stamping of this operation and the creation of a new version of the affected extension. In essence, these mechanisms should allow the creation of findable versions of a data subset. In order to implement such mechanisms, all these time marked database operations should be kept in a persistent store in order to maintain their history with the original data subset values and an appropriate approach to data versioning should be adopted. Several approaches have been proposed in literature for the implementation of these mechanisms. In summary, upon issuing a database operation, at time ti, that modifies a database subset the following actions should be carried out: (i) the database operation is time stamped with ti; (ii) this operation is inserted and maintained in a persistent store; (iii) a new version of the affected data subset is created, i.e., a new extension of each view defined on the updated data set; (iv) each newly generated extension is also time-stamped with ti. Therefore, a view is associated with different data operations that have affected its extensions. Both view and operations are endowed with time stamps that indicate the time of the execution of the data operation, i.e., the time of the modification of the view extension. The natural ordering of the time stamps associated with the operations that affect the extension of a view induces a natural ordering of the extensions produced by these operations, so that an extension can be seen as a version of the extension that precedes it in that ordering. Therefore, the time stamped views constitute a directed acyclic graph (DAG) whose nodes are the views at different points of time [View1(ti), View1(ti+x), .., View1(ti+x+n)] and the links represent temporal relationships between these nodes. A DAG has a topological ordering, a sequence of the nodes such that every arc is directed from earlier to later in the sequence. In essence, a DAG represents the evolution of a data subset over time. Finding a particular extension of a database view, i.e., the version of its extension at time ti implies (i) to identify the view through its ID; (ii) to cross the acyclic graph of this view until reaching the view View1(ti); and (iii) to execute the database operation associated with the View1(ti). The data integration problem can be slightly refined with respect to the formulation given in Sect. 4 as follows:

Given a number of distributed heterogeneous and time varying database views, represented by directed acyclic graphs, link them on the basis of existing semantic and temporal relationships among them.

To implement this new paradigm of data integration, we propose to proceed according to the following two steps: The first step consists in creating on top of each database view schema a level of virtual description. This virtualization describes the intentional part of a data subset (its schema) in terms of the RDF data model. Therefore, all database view schemata are represented in the RDF model while their extensions, i.e., the data subsets defined by them remain expressed in the data models supported by the local data management systems. This level of virtualization permits the uniform representation of the different schemata of the database views. Therefore, it allows researchers to query the database views and access their extensions (data subsets) by using a unique query language like SPARQL. This frees the researchers from the need to know the different query languages supported by the local database management systems in order to query the single database views.

The second step consists in inserting the virtualized database views together with their IDs and metadata in domain-specific registries.

The third step consists in linking the several virtualized database view schemata, described in these registries, on the basis of existing semantic and temporal relationships between them. The linking operation is guided by the adopted integration logic and produces a linked database view space where thematic graphs can be designed. These graphs constitute patterns of data that transform data in knowledge.

Finally, researchers by using appropriate query languages, as for example SPARQL, can explore the linked space by following appropriate links.

7   Conclusions

This paper presents a vision and outlines a direction to be followed in order to solve the data integration problem in the new emerging scientific data context. We identified the challenges that must be faced in order to successfully implement the new big research data integration paradigm that can be formulated as follows: "Given a number of distributed heterogeneous and time varying data sets, link them on the basis of existing semantic and temporal relationships among them." The main challenges we identified are:

- making the schema of a database view citable;
- endowing the schema of a database view with an identifier, a time stamp and appropriate metadata;
- creating mechanisms that support the versioning of the data subsets defined by the schemata of database views;
- developing mappings that allow an RDF application to access data subsets of non-RDF database views without having the need of transforming the data subsets into RDF triples;
- introducing temporal information into RDF graphs in order to be able to query this type of information;
- adopting an appropriate integration logic that has to guide a search engine in the identification of semantic and/or temporal relationships between distributed worldwide database view schemata and creating f links among them;
- developing or using existing domain-specific vocabularies/ontologies to support the process of semantically and temporarily linkage of database views;

- developing or using existing Catalogues/Registries where database view schemata are published;
- developing of query languages or extension of existing ones that allow to traverse linked RDF graphs.

References

1. Bernstein, P.A., Haas, L.M.: Information integration in the enterprise. Commun. ACM **51**(9), 72–79 (2008)
2. Bizer, C., Seaborne, A.: D2RQ-treating non-RDF databases as virtual RDF graphs. In: Proceedings of the 3rd international semantic web conference (ISWC 2004), vol. 2004 (2004)
3. Brackett, M.H.: Data Resource Integration: Understanding and Resolving a Disparate Data Resource, vol. 2. Technics Publications, Denville (2012)
4. Buneman, P., Davidson, S., Frew, J.: Why data citation is a computational problem. Commun. ACM **59**(9), 50–57 (2016)
5. Chawathe, S., et al.: The TSIMMIS project: integration of heterogenous information sources (1994)
6. Council, N.R., et al.: Steps Toward Large-scale Data Integration in the Sciences: Summary of a Workshop. National Academies Press, Washington, D.C. (2010)
7. Daraio, C., et al.: Data integration for research and innovation policy: an ontology-based data management approach. Scientometrics **106**(2), 857–871 (2016)
8. Doan, A., Halevy, A.Y.: Semantic integration research in the database community: a brief survey. AI Mag. **26**(1), 83–83 (2005)
9. Dong, X.L., Srivastava, D.: Big data integration. In: 2013 IEEE 29th International Conference on Data Engineering (ICDE), pp. 1245–1248. IEEE (2013)
10. Guarino, N., Oberle, D., Staab, S.: What is an ontology? In: Staab, S., Studer, R. (eds.) Handbook on Ontologies. IHIS, pp. 1–17. Springer, Heidelberg (2009). https://doi.org/10.1007/978-3-540-92673-3_0
11. Gutierrez, C., Hurtado, C.A., Vaisman, A.: Introducing time into RDF. IEEE Trans. Knowl. Data Eng. **19**(2), 207–218 (2007)
12. Halevy, A., Rajaraman, A., Ordille, J.: Data integration: the teenage years. In: Proceedings of the 32nd International Conference on Very Large Data Bases, pp. 9–16. VLDB Endowment (2006)
13. Heath, T., Bizer, C.: Linked data: evolving the web into a global data space. Synth. Lect. Semant. Web: Theory Technol. **1**(1), 1–136 (2011)
14. Idreos, S., Papaemmanouil, O., Chaudhuri, S.: Overview of data exploration techniques. In: Proceedings of the 2015 ACM SIGMOD International Conference on Management of Data, pp. 277–281. ACM (2015)
15. Kitchin, R.: Big data, new epistemologies and paradigm shifts. Big Data Soc. **1**(1), 2053951714528481 (2014)
16. Koch, C.: Data integration against multiple evolving autonomous schemata. Ph.D. thesis, Vienna U (2001)
17. Lenzerini, M.: Data integration: a theoretical perspective. In: Proceedings of the Twenty-First ACM SIGMOD-SIGACT-SIGART Symposium on Principles of Database Systems, pp. 233–246. ACM (2002)
18. McBride, B.: The resource description framework (RDF) and its vocabulary description language RDFs. In: Staab, S., Studer, R. (eds.) Handbook on Ontologies, pp. 51–65. Springer, Heidelberg (2004). https://doi.org/10.1007/978-3-540-24750-0_3

19. Naumann, F., Bilke, A., Bleiholder, J., Weis, M.: Data fusion in three steps: resolving inconsistencies at schema-, tuple-, and value-level. IEEE Data Eng. Bull. **29**(2), 21–31 (2006)
20. Proll, S., Rauber, A.: Scalable data citation in dynamic, large databases: model and reference implementation. In: 2013 IEEE International Conference on Big Data, pp. 307–312. IEEE (2013)
21. Silvello, G.: Theory and practice of data citation. J. Assoc. Inf. Sci. Technol. **69**(1), 6–20 (2018)
22. Vassiliadis, P.: A survey of extract-transform-load technology. Int. J. Data Warehous. Min. (IJDWM) **5**(3), 1–27 (2009)
23. Wiederhold, G.: Interoperation, mediation and ontologies. In: FGCS Workshop on Heterogeneous Cooperative Knowledge-Bases (1994)
24. Ziegler, P., Dittrich, K.R.: Three decades of data inteceration — all problems solved? In: Jacquart, R. (ed.) Building the Information Society. IIFIP, vol. 156, pp. 3–12. Springer, Boston, MA (2004). https://doi.org/10.1007/978-1-4020-8157-6_1

Text and Document Management

Construction of an In-House Paper/Figure Database System Using Portable Document Format Files

Masaharu Yoshioka[1,2(✉)] and Shinjiro Hara[3]

[1] Graduate School of Information Science and Technology,
Hokkaido University, Sapporo, Japan
`yoshioka@ist.hokudai.ac.jp`
[2] GI-Core, Hokkaido University, Sapporo, Japan
[3] Research Center for Integrated Quantum Electronics, Hokkaido University,
Sapporo, Japan

Abstract. Several general-purpose databases of research papers are available, including ScienceDirect and Google Scholar. However, these systems may not include recent papers from workshops or conferences for special interest groups. Moreover, it can be helpful for researchers in a particular domain to analyze research papers using their own terminology. To support these researchers, we propose a new database system based on the information extracted from portable document format documents that enables annotation of terms via a terminology extraction system. In this paper, we evaluate our system using a use-case experiment and discuss the appropriateness of the system.

1 Introduction

Many papers are published daily in academic journals, conference proceedings, and other media [1]. It is becoming increasingly likely that researchers wishing to find relevant information from the volume of material available will have to use an electronic library such as Science Direct[1] and/or a research paper database such as Google Scholar[2]. However, because the coverage by such services is currently considered to be incomplete, new research results distributed as conference or workshop papers for a particular research domain are often unavailable through these services. Although most of these papers are usually published in portable document format (PDF), it is not easy for researchers to access relevant material without having an adequate database that enables the organization and management of PDF files.

Having recognized the demand for handling PDF research papers, tools are available to extract content from PDF documents. The usual approach is to start by extracting the textual content from the PDF using a tool such as pdftext.

---

[1] http://www.sciencedirect.com/.

[2] https://scholar.google.com/.

© Springer Nature Switzerland AG 2019
D. Kotzinos et al. (Eds.): ISIP 2018, CCIS 1040, pp. 41–52, 2019.
https://doi.org/10.1007/978-3-030-30284-9_3

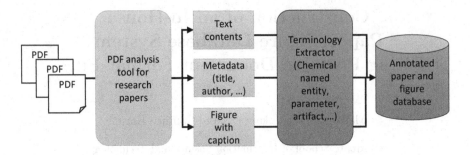

Fig. 1. Construction of an in-house database

In addition to text extraction, several additional tools have been proposed, including pdffigures2 [2][3], which is open-source software for extracting figure and table information with caption text as metadata.

In this paper, we describe a new framework for constructing an in-house paper/figure database in terms of a particular research domain using PDF documents that are collected by a research group. For example, most research groups would have CD-ROMS or USB sticks containing sets of proceedings from the conferences and workshops directly relevant to their research. In addition, they may have also collected PDF files of articles from journals and electronic libraries to which they subscribe. Our proposed framework uses such PDF files in the construction of an in-house paper/figure database. We also propose to use a terminology-extraction framework to construct a high-quality index for the database and for multifaceted analysis. Finally, we evaluate this framework by constructing a database for a nanocrystal-device development research group.

2   Database Framework

2.1   Construction of an In-House Database

Figure 1 shows the steps in the construction of an in-house database.

Extracting information from PDF files and storing them in a database involve two main steps, as follows.

1. Extraction of information from PDF files.
   Several tools can extract textual material and figures with captions from the papers. Based on the preliminary experiment of extracting textual material and figures with captions for nanocrystal device development papers, we selected pdffigures2 (see footnote 3), which was developed for Semantic Scholar[4] [2], one of the largest paper databases based on open-access papers.

---

[3] http://pdffigures2.allenai.org/.

[4] http://www.semanticscholar.org/.

2. Terminology extraction from textual material.

It is helpful for users to analyze papers with reference to their own terminology. An example is the normalization of terms represented in different formats, such as a pair comprising an acronym and its original description or a chemical entity expressed both as a formula and in words (e.g., CO_2 and carbon dioxide). Such an approach should significantly improve the recall of document-retrieval results. It is also helpful to construct a multifaceted analysis system (defined by terminology types) for the documents. This should be supported by domain-focused named-entity recognition systems (e.g., gene, protein, chemical, parameter) and terminology (ontology) construction tools [3,4].

2.2   Paper/Figure Retrieval System

The main retrieval functions of the proposed paper/figure retrieval system should be similar to those for conventional electronic libraries[5] and research paper databases. In addition to these functions, the system should implement a multifaceted analysis function using terminology types. Conventional electronic libraries such as ScienceDirect have features that enable multifaceted analyses of simple metadata in the research papers, such as journal name, year of publication, and author names. In addition, the user can specify the terminology types of interest. For example, if a user group is interested in chemical entities, the system can generate a list of such items, together with their document frequency. However, because of the size limitations of general databases, it may be preferable to utilize more specific databases for reference analyses.

3   Application to the Nanocrystal Device Development Domain

3.1   System Implementation

To evaluate the effectiveness of the proposed framework, we have implemented a system aimed at nanocrystal device development research groups.

Extraction of Information from PDF Files
Pdffigures2 (see footnote 3) is a tool used to extract textual material and figures with captions from the papers. It uses a research paper in PDF format as input and extracts the following three types of information from the PDF file.

- Image files for representing figures and tables in the paper.

---

[5] ScienceDirect previously had an image-retrieval feature using captions (https://www.elsevier.com/about/press-releases/science-and-technology/elsevier-releases-image-search-new-sciverse-sciencedirect-feature-enables-researchers-to-quickly-find-reliable-visual-content). However, this image-retrieval feature does not exist on the current version.

- Caption information of figures and tables in the paper (text, page, and reference to the extracted image file).
- Text content of the paper.

There are several cases where the tool fails to identify the positions of figures or tables. In such a case, no image file is generated. However, when the system succeeds to identify the caption part, caption text information is extracted without reference to the image file.

Terminology Dictionary Construction and Extraction

There are two main approaches to extract terminology from the text. One is a machine-learning based approach and the other is a dictionary-based approach. The machine-learning based approach is effective when large volumes of corpus are available for training. In such a case, the system can consider the context of the terminology for word sense disambiguation and develop a capability to identify newly introduced terminology. However, it is difficult to prepare such large volumes of corpus that represent the interests of small research groups who would be the typical target groups of the proposed system.

Therefore, we propose to use the dictionary-based approach for the database system. To extract terminology from the texts using the dictionary-based approach, it is necessary to have a framework to support such a terminology dictionary construction. In considering the construction of such a terminology dictionary, there are several research results for constructing such a dictionary from texts [4,5] and TermExtract [3][6] is an open source program based on such terminology research. We apply TermExtract to the texts that are extracted by pdffigures2, for generating terminology candidates. In addition, we also extract acronyms from the texts. This system extracts acronyms with a simple pattern-based tool that uses parentheses and regular expressions to identify pairs comprising an acronym and the original phrase. For example, acronyms are used to identify the same concept describing various surface descriptions and the distinction between chemical formulae and acronyms that use alphabetical symbols such as B, C, N, and W. Original forms of acronyms are also included in the terminology dictionary, even though their scores for the purpose of TermExtract are low.

In the last prototype system development, the user indicated that it is necessary to merge the different surface descriptions for the same concept (e.g., acronym (QD) and the original surface description (Quantum Dot)) [6]. Therefore, we use pairs of acronyms and original phrases to construct a dictionary for normalizing the surface descriptions in the papers. For example, corresponding original phrases of CMOS include "complementary metal oxide semiconductor," "complementary metal-oxide semiconductor," and "complementary MOS." Aggregated results based on one acronym and corresponding phrases are used for selecting one normalized form for representing the concept. If all corresponding surface descriptions represent the same concept, acronyms are used for the

---

[6] http://gensen.dl.itc.u-tokyo.ac.jp/termextract.html.

normalization. When there are two or more concepts that correspond to one acronym, representative phrases are used for normalization.

After selecting terminology candidates, we ask the users to classify the terms into meaningful categories. In this experiment, we use categories "artifact," "characteristic," "composite," "method," and "parameter." Since category of compound nouns can be identified using a category of head noun (e.g., category of "melting temperature" (parameter) is same as head noun "temperature" (parameter)), we first check the category of single noun terms that are frequently used as head of noun compounds and add tentative categories based on the head noun information.

In addition, since dictionary-based terminology extraction may generate inappropriate extraction results, we use the corpus annotation tool Brat[7] to edit the extracted results. The feedback results will be used for corpus construction for machine learning-based terminology extraction tools.

Chemical Named Entity Extraction

Since there are so many variations of chemical named entities, it is difficult to prepare a dictionary that covers all such variations. Therefore, we use machine-learning based chemical named entity recognition (CNER) tools to identify chemical named entities.

Because most CNER tools such as ChemSpot [7] and OSCAR [8] are optimized for biochemical-domain documents, they are not appropriate for our task. For example, a machine learning-based tool such as ChemSpot will tend to miss chemical named entities that rarely appear in biomedical documents. Rule-based systems such as OSCAR will tend to extract non-chemical named entities because they use patterns that represent organic chemical entities. Therefore, we used our own CNER tool based on conditional random field (CRF) [9], which employs surface descriptions, normalized parts of speech, orthogonal features, a chemical formula checker based on regular expressions, and comparisons with the entries in ChemIDPlus[8]. We trained the model using the CHEMDNER corpus [10] and the NaDev corpus [11].

3.2   Data Setup

We implemented the system using PDF files obtained from the journals and proceedings listed in Table 1, which cover topics related to nanocrystal device development. The table gives the number of articles, figure captions, and figures corresponding to captions extracted from the PDF files using pdffigures2. This tool has an option that enables captions to be extracted even if the system cannot extract the corresponding figure images from the PDF. Because pdffigures2 can use strong clues to identify captions, such as text strings that start with "Figure" or "Fig." followed by numbers, caption identification is usually straightforward. Therefore, we can specify the number of captions (independently of the figure images themselves) as shown in Table 1.

---

[7] http://brat.nlplab.org/.

[8] https://chem.nlm.nih.gov/chemidplus/chemidlite.jsp.

Table 1. Number of Articles, Captions, and Extracted Figures for Captions

Journal/Proceedings	Articles	Captions	Extracted Figures
Journal of Crystal Growth	88	588	569 (0.968)
Proceedings of SPIE	997	7,282	6,843 (0.940)
Japanese Journal of Applied Physics	875	6,505	6,312 (0.970)

In addition, pdffigures2 can successfully identify the corresponding figure images for most of the captions. However, the identification of the boundaries for the captions and/or the identification of the figure image area may be erroneous. In our present implementation, it is difficult to check the quality of all extracted images and captions, but we can record the proportion of figures for which an image area corresponding to each caption could be extracted. These proportions are shown as the bracketed numbers in Table 1. Although they may include unnecessary header text or may miss parts of the figure edges, most figures contain adequate information. Moreover, even for the failed cases, the system can identify the page numbers of the captions, which would enable the whole page to be specified as the "figure image" (this feature has yet to be implemented).

The following is categorical terminology information extracted using the system. These categories are selected based on the categories proposed in [11]. However, since dictionary-based matching cannot identify the role of terms (e.g., distinction between evaluation parameter and experiment parameter), several categories were merged for this experiment.

- Extraction by CNER tool.
  Chemical: Chemical named entity such as name of atom and molecule.
- Extraction using dictionary-based matching
  Material Characteristics: Characteristic feature of the material; e.g., magnetic or crystal.
  Composites: Name of material in particular style or constructed structure; e.g., film or particle.
  Artifact: Name of artifact used as final product; e.g., capacitor or solar cell.
  Parameter: Name of parameter for experiments, evaluations, and test device settings; e.g., temperature or frequency.
  Method: Name of parameter for experiments, evaluations, and tests; e.g., fabrication or scanning electron microscope.

Using these categories, we can support structured queries that are constructed by users to describe their information needs. For example, suppose a user has an interest in the electric current (a parameter) of a solar cell (an artifact). The user would construct a query using the name of the target artifact (solar cell) and one of its parameters (electric current). The figures retrieved for this query may contain various types of graphs for evaluating the electric current of a solar cell. The counting frequency of parameter names in the captions may suggest relevant parameters for use in controlling the target parameter (electric current). The user can then select a parameter of interest, and the system will then present the varieties of graphs that describe relationships between the target parameter and the selected parameter. A comparison between these graphs may be helpful in identifying common tendencies among these parameters. In addition, if there are graphs that differ from a common tendency, expert users can use their expertise to identify such graphs among the results.

3.3 Multifaceted Figure Retrieval System

We constructed a database using the information extracted from the PDF files and used the database for a multifaceted figure retrieval system [6]. The system shows the top 10 most-frequently used terms for each category in the retrieved results, which aims to provide an understanding of the characteristic terms in the results. Figure 2 shows a screenshot of the system for retrieving information of the figures using the "thickness" parameter (i.e., corresponding figure images are excluded because of copyright limitations). By inspecting the top 10 keywords listed, the user can revise the query to narrow down the results, by selecting different keywords. Figure 3 shows a result of the system for the revised query (e.g., selecting "film" from the "composites" keyword list).

3.4 Use-Case Experiment

To evaluate the appropriateness of the system, we conducted a use-case experiment with two Master's students from our nanocrystal research laboratory, which collects relevant research papers in PDF format. The experiment included instruction in using the system, checking the quality of the terminology dictionary, checking the terminology extraction results, and trial use of the system with respect to the participants' own research interests.

The experiment was scheduled over four days (four hours per day), as follows:

1st day: general instruction
2nd day: checking the terminology dictionary used for extraction
3rd day: general instruction in using the corpus annotation tool to revise the term extraction errors
4th day: trial use of the system with individual research interests.

After the final trial, we conducted a user survey to evaluate the system. Table 2 shows the results of the survey, which were graded from 1 (bad) to 5 (good).

In addition to the user-grading evaluation, we asked the following questions:

Which functions of the system are useful?
- Multifaceted analysis
  The facet–keyword list of the retrieved results is useful (e.g., identify a chemical material list for the query "magnetic")
- List of figures and captions
  Figures are a good way to index potentially useful papers.

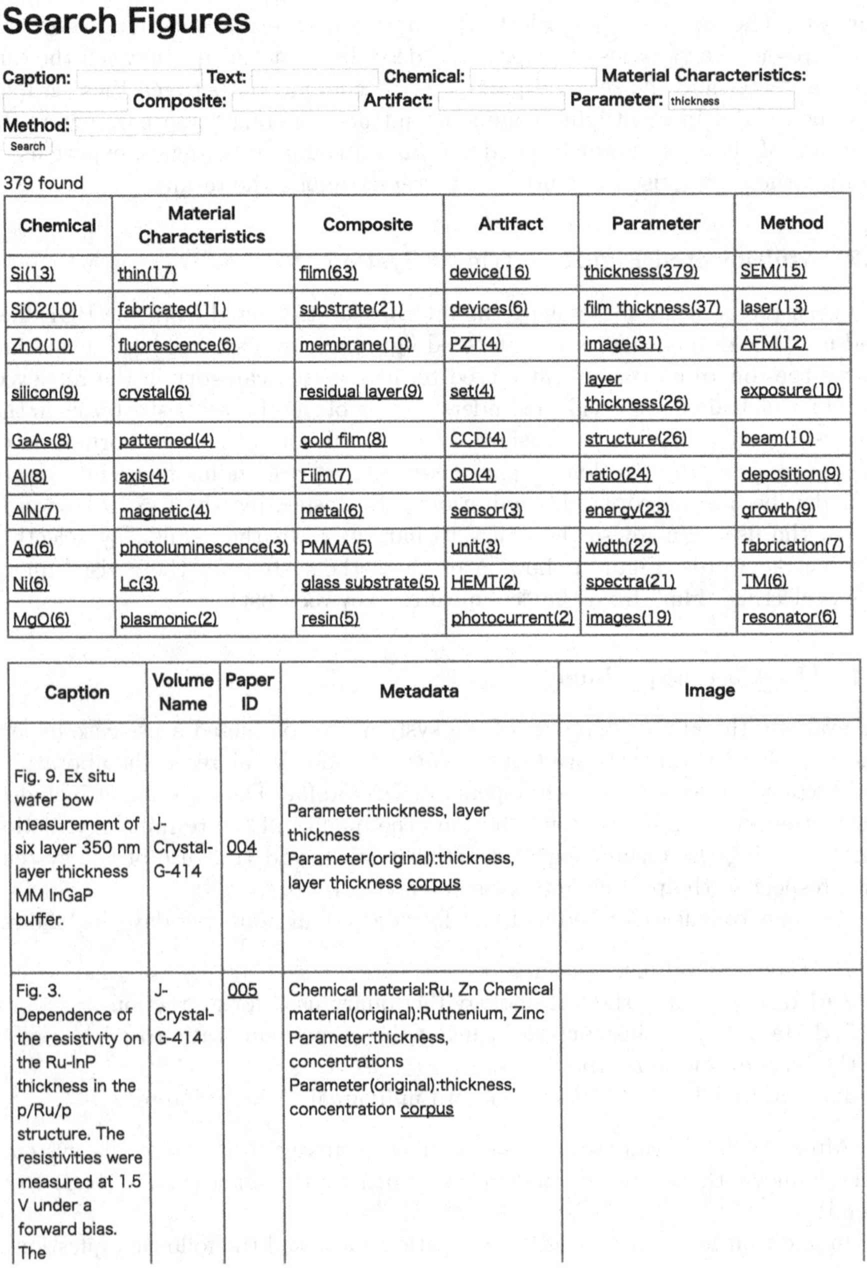

Search Figures

Caption:		Text:		Chemical:		Material Characteristics:
	Composite:		Artifact:		Parameter: thickness	

Method:

Search

379 found

Chemical	Material Characteristics	Composite	Artifact	Parameter	Method
Si(13)	thin(17)	film(63)	device(16)	thickness(379)	SEM(15)
SiO2(10)	fabricated(11)	substrate(21)	devices(6)	film thickness(37)	laser(13)
ZnO(10)	fluorescence(6)	membrane(10)	PZT(4)	image(31)	AFM(12)
silicon(9)	crystal(6)	residual layer(9)	set(4)	layer thickness(26)	exposure(10)
GaAs(8)	patterned(4)	gold film(8)	CCD(4)	structure(26)	beam(10)
Al(8)	axis(4)	Film(7)	QD(4)	ratio(24)	deposition(9)
AlN(7)	magnetic(4)	metal(6)	sensor(3)	energy(23)	growth(9)
Ag(6)	photoluminescence(3)	PMMA(5)	unit(3)	width(22)	fabrication(7)
Ni(6)	Lc(3)	glass substrate(5)	HEMT(2)	spectra(21)	TM(6)
MgO(6)	plasmonic(2)	resin(5)	photocurrent(2)	images(19)	resonator(6)

Caption	Volume Name	Paper ID	Metadata	Image
Fig. 9. Ex situ wafer bow measurement of six layer 350 nm layer thickness MM InGaP buffer.	J-Crystal-G-414	004	Parameter:thickness, layer thickness Parameter(original):thickness, layer thickness corpus	
Fig. 3. Dependence of the resistivity on the Ru-InP thickness in the p/Ru/p structure. The resistivities were measured at 1.5 V under a forward bias. The	J-Crystal-G-414	005	Chemical material:Ru, Zn Chemical material(original):Ruthenium, Zinc Parameter:thickness, concentrations Parameter(original):thickness, concentration corpus	

Fig. 2. Screenshot of the figure retrieval system (initial query)

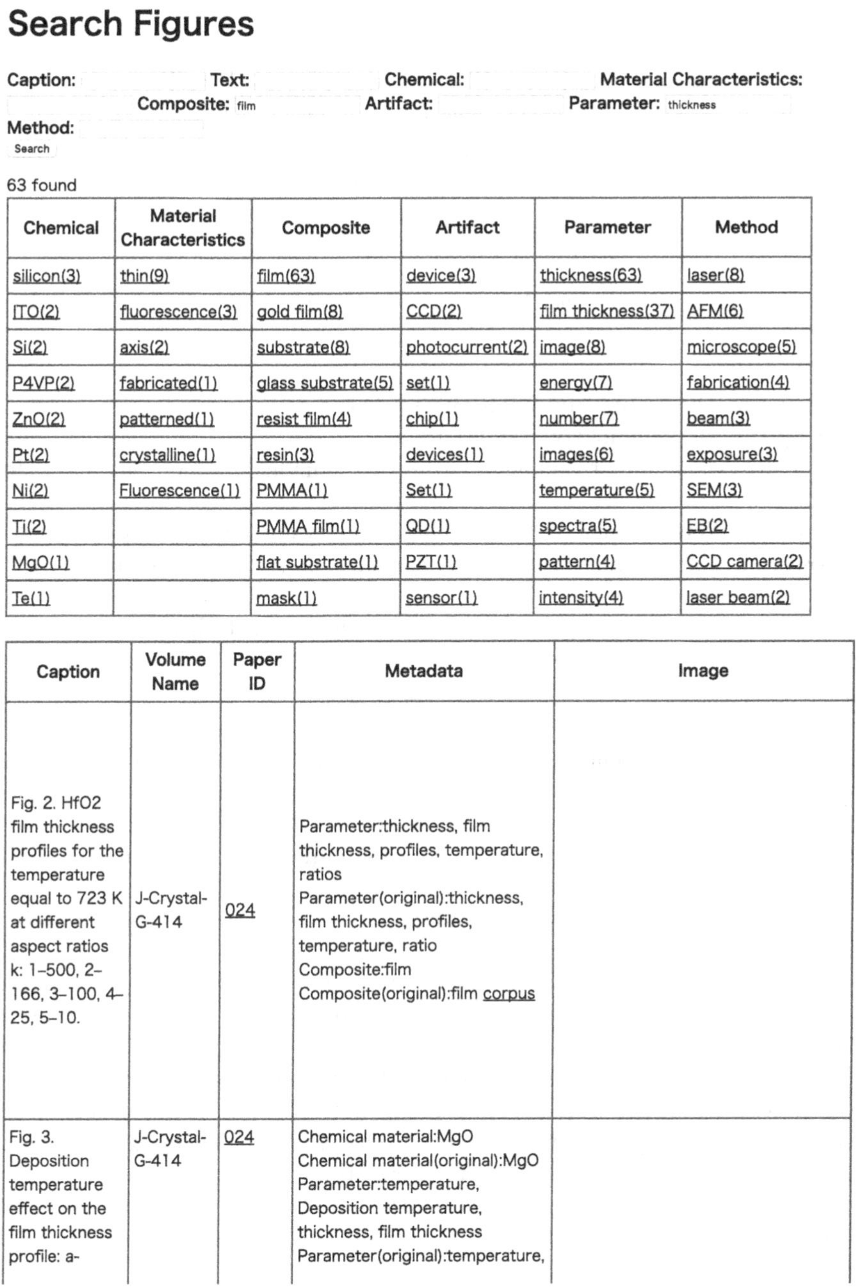

Search Figures

Caption:	Text:	Chemical:	Material Characteristics:
	Composite: film	Artifact:	Parameter: thickness

Method:
Search

63 found

Chemical	Material Characteristics	Composite	Artifact	Parameter	Method
silicon(3)	thin(9)	film(63)	device(3)	thickness(63)	laser(8)
ITO(2)	fluorescence(3)	gold film(8)	CCD(2)	film thickness(37)	AFM(6)
Si(2)	axis(2)	substrate(8)	photocurrent(2)	image(8)	microscope(5)
P4VP(2)	fabricated(1)	glass substrate(5)	set(1)	energy(7)	fabrication(4)
ZnO(2)	patterned(1)	resist film(4)	chip(1)	number(7)	beam(3)
Pt(2)	crystalline(1)	resin(3)	devices(1)	images(6)	exposure(3)
Ni(2)	Fluorescence(1)	PMMA(1)	Set(1)	temperature(5)	SEM(3)
Ti(2)		PMMA film(1)	QD(1)	spectra(5)	EB(2)
MgO(1)		flat substrate(1)	PZT(1)	pattern(4)	CCD camera(2)
Te(1)		mask(1)	sensor(1)	intensity(4)	laser beam(2)

Caption	Volume Name	Paper ID	Metadata	Image
Fig. 2. HfO2 film thickness profiles for the temperature equal to 723 K at different aspect ratios k: 1–500, 2–166, 3–100, 4–25, 5–10.	J-Crystal-G-414	024	Parameter:thickness, film thickness, profiles, temperature, ratios Parameter(original):thickness, film thickness, profiles, temperature, ratio Composite:film Composite(original):film corpus	
Fig. 3. Deposition temperature effect on the film thickness profile: a–	J-Crystal-G-414	024	Chemical material:MgO Chemical material(original):MgO Parameter:temperature, Deposition temperature, thickness, film thickness Parameter(original):temperature,	

Fig. 3. Screenshot of the figure retrieval system (revised query)

Table 2. User survey results

Question	User 1	User 2
Coverage of research papers	2	2
Coverage of research areas	3	4
Quality of caption-extraction results	3	3
Quality of term extraction	2	2
System interface usability	3	3
System response time	4	2

What are the problems with the system?
- Quality of terminology extraction
  The system cannot handle varieties of synonyms and concept hierarchy terms.
  (e.g., different usage of "–" or "GaN layer" vs "GaN nucleation layer")

How many relevant articles did you find using this database?
- User 1: five papers (of which three papers were new to him)
- User 2: nine papers (of which six papers were new to her)

3.5 Discussion

We understand that a quantitative analysis obtained from only two participants in the experiment is not sufficiently meaningful. However, since the two participants evaluated the system from the perspective of daily usage in their research activities, the issues that they identified are useful in giving direction to improving the system.

From the user survey, we found that construction of in-house paper databases may adequately cover specific research areas (grades for the coverage of research areas were 3 and 4). However, it would appear necessary to expand the paper database to cover more relevant papers (grades for the coverage of research areas were both 2).

In checking the terminology dictionary for its extraction performance, the users indicated that the dictionary was inadequate. Because of this problem, the grading for term extraction was also poor (grades for the coverage of research areas were both 2). In addition, the users would prefer the term extraction results to be aggregated using synonyms and concept hierarchies. Therefore, it would be necessary to provide a framework to support such a dictionary-maintenance process.

With respect to caption extraction, there were several cases for which the system failed to extract appropriate captions. Although there were fewer failures than successes, the grades for caption extraction were not high (both 3). This may suggest that users would require near-perfect extraction results if a grading of 5 was to be given.

In their overall evaluation of the system, the users found the main functionality of the system (the multifaceted analysis results and list of figures) to be useful. However, the usability and response time of the prototype system is inadequate (usability grades were both 3 and the responsiveness grades were 4 and 2, respectively). It will be necessary to rework aspects of the system to improve the quality of the user experience.

Since it is not a simplistic process to set up appropriate research papers in PDF format for a particular research domain and recruiting researchers to evaluate such a prototype system, the number of participants of the experiments are limited at this stage of the research. However, the issues raised by the two participants in the experiment described in this paper are useful for the further development and refinement of the system.

Although the above evaluation of the system was based on the experience of two participants from a specific interest field searching for papers, it only partially reflects the needs of potential users of the system, and it remains necessary to evaluate the system with researchers from other interest groups.

4   Related Work

There have been several attempts to construct paper/figure databases via extraction from open-access journals.

Semantic Scholar (see footnote 4) [2] is an academic search engine that constructs databases using open-access papers distributed as PDF files. This system can also search figures via caption information. However, it has no tool to support interactive query construction. NOA [12] is a figure database constructed from open-access journal information provided in HTML format. One of the main features of this system is its incorporation of two items of metadata derived from machine-learning results, i.e., figure types (e.g., photo or diagram) and content tags (e.g., "materials science" or "thermodynamics"). These metadata items may be helpful in selecting relevant papers from a wide variety of research fields, but there is no specific tool for refining queries to focus on papers that match the researcher's particular interests.

5   Conclusions

In this paper, we have introduced a new in-house paper/figure database system for particular research domains. The system extracts information from PDF files for journal papers and/or proceedings collected by a particular research group. We conducted use-case experiments with a prototype system involving researchers from the nanocrystal device development domain.

From our experiments, we confirmed that the system functionality implemented on the prototype system seemed to be useful for the researchers, but it will be necessary to improve the quality of terminology extraction and system usability for the system to be deployed in a practical research environment. Future work will also need to include finding several appropriate research groups to use and evaluate this system so that the quality of the evaluation results can be improved.

Acknowledgments. This research was partly supported by ROIS NII Open Collaborative Research 2018-24.

References

1. Ware, M., Mabe, M.: The STM report: an overview of scientific and scholarly journal publishing, International Association of Scientific, Technical and Medical Publishers (2015). http://www.stm-assoc.org/2015_02_20_STM_Report_2015.pdf
2. Clark, C., Divvala, S.: PDFFigures 2.0: mining figures from research papers. In: 2016 IEEE/ACM Joint Conference on Digital Libraries (JCDL), pp. 143–152 (2016)
3. Nakagawa, H., Mori, T.: A simple but powerful automatic term extraction method. In: COLING-02 on COMPUTERM 2002: Second International Workshop on Computational Terminology - Volume 14. COMPUTERM 2002, Stroudsburg, PA, USA, pp. 1–7, Association for Computational Linguistics (2002)
4. Blaschke, C., Valencia, A.: Automatic ontology construction from the literature. Genome Inf. **13**, 201–213 (2002)
5. Kageura, K., Yoshioka, M., Koyama, T., Nozue, T., Tsuji, K.: Towards a common testbed for corpus-based computational terminology. In: Computerm 1998, pp. 81–85 (1998)
6. Yoshioka, M., Zhu, T., Hara, S.: A multi-faceted figure retrieval system from research papers for supporting nano-crystal device development researchers. In: The Proceedings of the First International Workshop on Scientific Document Analysis (SCIDOCA 2016), The Japanese Society of Artificial Intelligence (2016). Short paper 2
7. Rocktäschel, T., Weidlich, M., Leser, U.: ChemSpot: a hybrid system for chemical named entity recognition. Bioinformatics **28**, 1633–1640 (2012)
8. Jessop, D., Adams, S., Willighagen, E., Hawizy, L., Murray-Rust, P.: OSCAR4: a flexible architecture for chemical text-mining. J. Cheminform. **3**, 41 (2011)
9. Dieb, T.M., Yoshioka, M.: Extraction of chemical and drug named entities by ensemble learning using chemical ner tools based on different extraction guidelines. Trans. Mach. Learn. Data Min. **8**, 61–76 (2015)
10. Krallinger, M., et al.: The chemdner corpus of chemicals and drugs and its annotation principles. J. Cheminform. **7**, S2 (2015)
11. Dieb, T.M., Yoshioka, M., Hara, S.: An annotated corpus to support information extraction from research papers on nanocrystal devices. J. Inf. Process. **24**, 554–564 (2016)
12. Charbonnier, J., Sohmen, L., Rothman, J., Rohden, B., Wartena, C.: NOA: a search engine for reusable scientific images beyond the life sciences. In: Pasi, G., Piwowarski, B., Azzopardi, L., Hanbury, A. (eds.) ECIR 2018. LNCS, vol. 10772, pp. 797–800. Springer, Cham (2018). https://doi.org/10.1007/978-3-319-76941-7_78

A Domain-Independent Ontology for Capturing Scientific Experiments

Zakariae Aloulen[1,2]($\boxtimes$), Khalid Belhajjame[1], Daniela Grigori[1], and Renaud Acker[2]

[1] Paris-Dauphine University, PSL Research University, CNRS, [UMR 7243], LAMSADE, 75016 Paris, France
{zakariae.aloulen,khalid.belhajjame,daniela.grigori}@dauphine.fr
[2] AgiLab Company, 75019 Paris, France
{zakariae.aloulen,renaud.acker}@agilab.com

Abstract. Semantic web technologies have proved their usefulness in facilitating the documentation, annotation and to a certain extent the reuse and reproducibility of scientific experiments in laboratories. While useful, existing solutions suffer from some limitations when it comes to supporting scientists. Indeed, it is up to him/her to identify which ontologies to use, which fragments of those ontologies are useful for his experiments and how to combine them. Furthermore, the behavior and constraints of the domain of interest to the scientist, e.g., constraints and business rules, are not captured and as such are decoupled from the ontologies. To overcome the above limitations, we propose in this paper a methodology and underlying ontologies and solutions with the view to facilitate for the scientist the task of creating an ontology that captures the specificities of the domain of interest by combining existing well-known ontologies. Moreover, we provide the scientist with the means of specifying behavioral constraints, such as integrity constraints and business rules, with the ontology specified. We showcase our solution using a real-world case study from the field of agronomy.

Keywords: Core ontology · Domain ontology ·
Laboratory Information Management System · Scientific experiments ·
Business rules

1 Introduction

For several years, the volumes of data available in research and development (R&D) laboratories have grown exponentially. These data are produced by laboratory instruments, extracted from external data sources, and generated by partner laboratories, among other sources.

Based on a laboratory information management system (LIMS), R&D laboratories perform a set of key operations on the generated data. A LIMS is a software framework that provides support for tracking data and workflows (experiments), data reporting, data exchange interfaces, etc.

© Springer Nature Switzerland AG 2019
D. Kotzinos et al. (Eds.): ISIP 2018, CCIS 1040, pp. 53–68, 2019.
https://doi.org/10.1007/978-3-030-30284-9_4

The interpretation of research data depends heavily on the use of metadata [25]. With the emergence of semantic web standards, new methods and models for scientific data exchange, reproducibility and annotation have been developed [3,4,16]. Some of these models, see e.g., [8], provide ontologies to annotate scientific experiments stored in a LIMS. However, such ontologies are still limited as the defined shared vocabulary is domain-dependent and useful only in particular laboratory use cases. Other approaches, see e.g., [9], generate models that are specific to a scientific project and its technical constraints and business rules. Such models remain still inadequate for scientists as they are difficult to transfer or reuse in other fields. Indeed, it is up to the scientist to identify which ontologies to use, which fragments of those ontologies are useful for his experiments and how to combine them.

Because of this, scientists often end-up creating their new ontology that is not aligned with existing ontologies, thereby reducing the benefit that can be harvested from the use of linked data technologies. Furthermore, the behavior and constraints of the scientist's application domain, e.g., constraints and business rules, are not necessarily captured and as such are decoupled from the ontologies.

To overcome the above limitations, we propose in this paper a methodology and underlying ontologies that facilitate the task of creating a domain ontology by combining existing well-known ontologies. Moreover, we provide the scientist with the means of specifying and associating constraints, such as integrity constraints and business rules, with the ontology specified.

Specifically, we make the following contributions:

- A *domain-independent methodology* for composing the ontology that captures the domain of interest to the scientist (Sect. 2).
- A *Scientific Experiments Core Ontology (SECO)*. We specify the ontologies that we selected to create a new core ontology for annotating scientific data produced in laboratories with the flexibility to be extended to different application domains (Sect. 3). We also specify the requirements that we elicited and that guided us in our choices.
- *Case-study*. We showcase our solution using a real-world case study from the field of agronomy (Sect. 4).

Moreover, we analyze and compare related work to our proposal (Sect. 5), and conclude the paper by highlighting our contributions and discussing future work (Sect. 6).

2   A Generic Methodology for Domain-Specific Requirements

There exist a plethora of solutions [3,4,16] that have been developed since the inception of the linked data to design ontologies for the annotation and capture of scientific experiments' knowledge. While useful, these ontologies tend to be domain-specific and as such are not amenable for cross-domains utilization. For example, one could use the Allotrope ontology as it is, however, such ontologies mix up domain-independent and domain-dependent concepts, making it difficult for a prospective user to tease them apart. Another alternative, would be to use for each field a domain-ontology at its whole, however, such an approach is likely yield an ontology with concepts that are of no interest to the user, and will still need to be extended with new concepts and properties. The approach we opted for is therefore to have a small core ontology that can be easily extended to meet the requirements of each domain/scientist.

To overcome this issue, we designed a methodology (in Fig. 1) that is generic in the sense that it can be used for modeling (or more accurately composing) an ontology for capturing scientific experiments without compromising the peculiarities of the domain of interest.

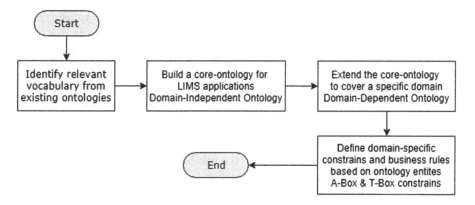

Fig. 1. A domain-independent methodology for capturing scientific experiments' knowledge

The first step consists in identifying vocabulary from existing ontologies. The reused vocabulary must align with that of the LIMS scientific workflow metamodel (in Fig. 2).

The second step is building a core ontology based on existing ontologies. Such an ontology contains domain-independent concepts and properties that can be used for the description of the main characteristics of a scientific experiment, for example, its analysis specification and execution, the parameters' results of a sample analyzed in this experiment and the limits that some parameters must not exceed.

The third step consists in extending the core-ontology to cover a domain of interest, e.g., agriculture, life sciences, etc. Often the use of an existing ontology is not sufficient since the scientist would want to model aspects that are particular to the scientific experiments' domain.

Finally, the scientist may want to add domain-specific constraints and business rules. These constraints may be expressed in terms of the ontology (T-Box) or the instances of those terms (A-Box). Specifically:

- Terminological knowledge (T-Box). Refers to statements that describe the hierarchy of concepts, and the relationships between them.
- Assertional knowledge (A-Box). Refers to ontology instances (individuals) which is more specific and more tied to a specific context.

These two components are combined to specify integrity constraints or used to assert object properties' relationships and values of certain data properties using user-defined functions.

For example, a sample which was sampled from a specific type of soil, have to be analyzed within a specific type of analysis and/or parameters (T-Box constraint) and/or for some parameters, the values must not exceed certain limits (A-Box constraint).

A business rule is a set of statements that constrain or define some aspects of a domain of interest. For example, a sample that has a mercury value greater than a certain threshold must be banned for human consumption. In our case, it is constructed by a combination of A-Box and/or T-Box restrictions.

The ontology and constraints obtained as a result are then annotated and published within the Open Biological and Biomedical Ontologies (OBO) Foundry [19] registry to be eventually (re)-used by other scientists.

While conceptually interesting and domain-independent, the methodology defined above raises some challenges. Perhaps the most important one is the choice of the ontologies used in the steps (1) and (3). In fact, there isn't a single ontology that is readily available to play the role of the core ontology. Instead, we have to compose such an ontology by re-using fragments of multiple ontologies. As we will show in the next section, we adopted a systematic approach guided by a set of requirements that we elicited for this purpose.

3   SECO: Scientific Experiments Core Ontology

The main goal of our work is to model a generic ontology (steps (1) and (2) in Fig. 2) that satisfies the following prerequisites:

- Allow annotating the scientific data stored in a Laboratory Information Management System (LIMS).
- Can be customized to potentially any application domain.

To reach our objectives, we start by identifying the key requirements that need to be satisfied by our core ontology. We then go on to specify how we constructed a core ontology that meets the requirements listed and that is obtained by composing fragments from existing ontologies.

Requirement 1. The core ontology needs to provide scientists with the means to capture information about their scientific experiments stored in a LIMS.

LIMS is a computer system for laboratory data management that enables laboratories to trace their activities such as analysis of samples, used instruments, inventory levels and users. Figure 2 represents a metamodel of the scientific workflow used in a LIMS. It consists mainly of the following entities:

- A Sample is received by a laboratory and analyzed following a Request (from the same or external laboratory).
- A Test (the instance of Analysis) is performed on a given a sample based on several Parameters (analysis' parameters). These parameters must not exceed certain Limits.
- Depending on the method and procedure, the Analysis is prepared by the Analyst using the required Reagents and performed on the measuring Instruments.

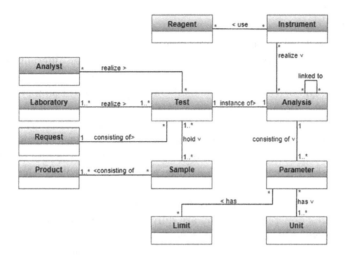

Fig. 2. Scientific workflow metamodel of a Laboratory Information Management System.

Requirement 2. The main entities of the LIMS' scientific workflow must be domain-independent. The annotation of this entities and relationships between them, must be decoupled from the application domain of the scientist and its specific constraints.

To respond to this requirement, we reuse fragments from the Allotrope Foundation Ontologies[1] (AFO). The main advantage of AFO is that it is based on laboratories' needs, their scientific and measurements workflows. However, AFO is still limited as :

- It defines terms for R&D laboratories in a general way based on some case studies (e.g. biological, chromatography, chemical experiments).
- The constraints and relationships between entities are not explicitly asserted. In our case, we must define constraints and links (Object and Datatype Properties) between the main entities of the LIMS scientific workflow.
- The vocabulary of AFO is not sufficient for the annotation of all the terms used in the LIMS core-workflow and/or other laboratory domains of expertise.

---

[1] AFO Ontologies: https://www.allotrope.org/ontologies.

To overcome these limitations, we draw upon AFO vocabulary, and other core and domain ontologies:

- Friend of a Friend[2] (FOAF) and vCard[3] ontologies to describe persons and organizations and their activities (Analyst, Laboratory, Address etc.).
- Relation Ontology[4] (RO) to describe relationships between entities.
- Environment Ontology[5] (ENVO) to describe the environment entities related to the sample of a scientific experiment.

These ontologies complete the vocabulary reused from AFO, as they describe entities that can be used to cover standards and requirements that must be involved in a LIMS.

Requirement 3. The entities of the core-ontology must be easily extended to include additional specific classes and/or properties to cover different application domains (e.g., agronomy, pharmacy, geology). For this purpose, we use ontologies from the Open Biological and Biomedical Ontologies (OBO) Foundry.[6] The OBO Foundry is a collaborative experiment in which multiple ontologies describing several domains of expertise are involved. These ontologies working together within the framework of the OBO Foundry utilize the Basic Formal Ontology (BFO) [1] as the starting point for the categorization of entities and relationships.

BFO is used by scientists to provide a top-level ontology that can serve as a common starting point for the creation of core and domain ontologies in different areas of science. It provides a formal-ontological hierarchy and a set of general terms and relations that are currently being used in a large wide of domains. The core-ontology we propose in this paper use BFO as an upper-level ontology, which may facilitate the reuse of fragments from OBO ontologies to cover different scientific experiments' fields.

Based on the metamodel of the LIMS' scientific workflow (in Fig. 2) and the specified requirements, we composed the SECO ontology, which is illustrated in Fig. 3. SECO reuse fragments from the ontologies stated above, and is made up of the following four main entities:

- **An analysis definition** (obi:plan corresponds to Analysis entity) represents the analysis performed in a laboratory (obib:laboratory corresponds to Laboratory entity) and consists of one or more parameters (e.g., mass, viscosity of a sample, concentration of a component).
- **An analysis instance** (af-p:test corresponds to Test entity) represents the execution of an analysis. It is performed on one or more samples and generates some results (af-r: scalar datum).
- **A sample** (af-m:sample corresponds to Sample entity) is collected from a location (envo:environmental feature) and analyzed in an analysis process (af-p:test).

---

2 FOAF Ontology : http://www.foaf-project.org.
3 vCard Ontology: https://www.w3.org/TR/vcard-rdf/.
4 RO: http://www.obofoundry.org/ontology/ro.html.
5 ENVO Ontology: https://bioportal.bioontology.org/ontologies/ENVO.
6 OBO Foundry: http://www.obofoundry.org.

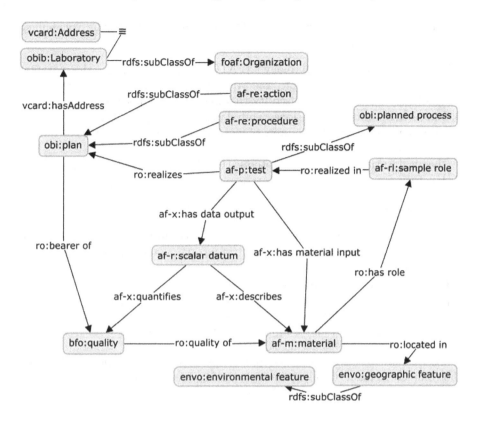

Fig. 3. Scientific Experiments Core Ontology.

– **A parameter** (bfo:quality corresponds to Parameter entity) is quantified by
  a value (af-r:scalar datum) and qualifies a physical entity (af-m:material), for
  example, a sample.

SECO ontology is publicly available online[7] under the Creative Commons
Attribution 4.0 International License[8].

The proposed core ontology may be used and extended by R&D laboratories
from different application domains. In the next section, we will illustrate how
the SECO vocabulary can be specialized to cover the agronomy domain.

4   Case Study

In order to validate our approach, we have defined an ontology for the agronomy
domain. To define the model, we relied on a LIMS database that stores data
about scientific experiments' results in the field of agronomy. For the sake of
simplicity, we present here a reduced version of the main model (Fig. 4).

---

[7] Scientific Experiments Core Ontology: https://github.com/aloulen/SECO.
[8] CC BY 4.0: http://creativecommons.org/licenses/by/4.0/.

The model extends the SECO ontology (see Sect. 3) by reusing vocabulary from the following ontologies:

- Phenotype And Trait Ontology[9] (PATO) used to describe the parameters of a sample, for example, the volume (pato:volume), the temperature (pato:temperature) or the viscosity (pato:viscosity) of a sample.
- Quantities, Units, Dimensions, and Types Ontology[10] (QUDT) used to describe physical quantities, units of measure, and their dimensions in various measurement systems.
- Chemical Entities of Biological Interest (ChEBI) [6] to describe some chemical structures.

The Agronomy Ontology consists of three main entities:

- **Analysis.** The sample (af-m:sample) is raised from a sampling location (envo:garden) and will be analyzed (af-p:test) based on some parameters (bfo:quality).
- **Parameters.** Each sample is qualified by some parameters (in this case, pato:pressure, pato:temperature and pato:mass) that must be quantified.
- **Results.** The quantified results (af-r:scalar quantity datum) are characterized by a unit (qudt:unit) and a value (qudt:numeric value).

Figure 5 illustrates a partial, simplified instantiation of the proposed model. The complete version of the agronomy ontology is available online[11].

To showcase the feasibility and usefulness of our approach, we have followed the steps below:

- The first step consisted in annotating scientific experiments data stored in the LIMS of the agronomy laboratory of Paris city hall, provided by AgiLab[12] company.
  The LIMS of AgiLab company use a relational database (RDB) to store scientific experiments' data. To be able to use our solution, we created mappings between the RDB and the ontology vocabulary using RDB to RDF Mapping Language (R2RML) [5]. The mappings are then used to produce the corresponding RDF (Resource Description Framework) [15] statements.
- In the second step, we have translated some expert domain questions to the SPARQL Protocol and RDF Query Language (SPARQL) [13]. Listings 1.1 and 1.2 show the SPARQL queries that return responses to the following questions:

---

[9] PATO Ontology: http://www.obofoundry.org/ontology/pato.html.
[10] QUDT Ontologies: http://www.qudt.org/release2/qudt-catalog.html.
[11] Scientific Experiments Agronomy Ontology: https://github.com/aloulen/SECO_AGRO.
[12] AgiLab is a software development company that provides information systems (LIMS) for managing the activities of research and development laboratories in different domains of activity.

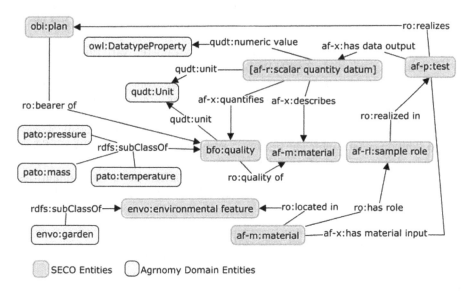

Fig. 4. Simplified ontology of agronomy scientific experiments.

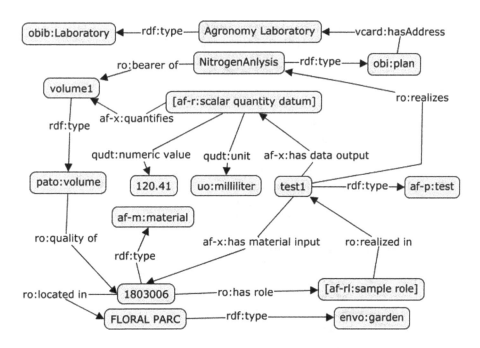

Fig. 5. Agronomy Ontology instantiation example.

- Are there any samples that have been raised from different locations and have had the same results under the same parameters ? (SPARQL Query 1)
- Between two given dates, what is the parameter variation (analysis' results) of a sample ? (SPARQL Query 2)

To make the SPARQL queries more visible for the reader, we have used the main label of each entity instead of its ID (for example "af-x:has value" instead of "af-x:AFX_0000690").

- In the final step, we have used the business rule below based on some agronomy domain constraints. The Semantic Web Rule Language (SWRL) [14] was used to define such constraints (in Listing 1.3):
  - If a sample raised from a given garden has a lead or mercury quantities greater than a certain threshold, than this garden must be placed under surveillance.

```
PREFIX xsd: <http://www.w3.org/2001/XMLSchema#>
PREFIX bfo: <http://purl.obolibrary.org/obo/BFO/>
PREFIX af-x: <http://purl.allotrope.org/ontologies/property#>
PREFIX af-m: <http://purl.allotrope.org/ontologies/material#>
PREFIX af-rl: <http://purl.allotrope.org/ontologies/role#>
PREFIX af-p: <http://purl.allotrope.org/ontologies/process#>
PREFIX af-r: <http://purl.allotrope.org/ontologies/result#>
PREFIX qudt: <http://qudt.org/schema/qudt#>
PREFIX ro: <http://purl.obolibrary.org/obo/>
PREFIX envo: <http://purl.obolibrary.org/obo/>

SELECT ?sample1 ?sample2 ?param1 ?result1 ?loc1 ?loc2
WHERE
{
    ?sample1        a af-m:material;
                    ro:has role ?samplerole1;
                    ro:realized in ?test2;
                    ro:located in ?loc1, ?loc2.
    ?sample2        a af-m:material;
                    ro:has role ?samplerole2;
                    ro:realized in ?test1;
                    ro:located in ?loc2.
    ?samplerole1    a af-rl:sample role.
    ?samplerole2    a af-rl:sample role.
    ?test1          a af-p:test;
                    af-x:has data output ?data;
                    ro:bearer of ?param1.
    ?test2          a af-p:test;
                    af-x:has data output ?data;
                    ro:bearer of ?param2.
```

```
    ?data1        a af-r:scalar quantity datum;
                  af-x:quantifies ?param1;
                  qudt:numeric value ?result1.
    ?data2        a af-r:scalar quantity datum;
                  af-x:quantifies ?param2.
                  qudt:numeric value ?result2.
    ?param1       a bfo:quality.
    ?param2       a bfo:quality.
    ?loc1         a envo:garden.
    ?loc2         a envo:garden.

    FILTER (?sample1 != sample2 && ?test1 = ?test2 &&
            ?param1 = ?param2 && ?loc1 != ?loc2
            && ?result1 = ?result2)
} LIMIT 100
```

Listing 1.1. SPARQL Query 1

```
SELECT ?sample ?result ?mtime ?loc
WHERE
{
    ?sample       a af-m:material;
                  ro:has role ?samplerole;
                  af-x:has measurement time ?mtime;
                  ro:realized in ?test;
                  ro:located in ?loc.
    ?samplerole   a af-rl:sample role.
    ?test         a af-p:test;
                  af-x:has value "NitrogenAnalysis"^^xsd:string;
                  af-x:has data output ?data;
                  ro:bearer of ?mass.
    ?data         a af-r:scalar quantity datum;
                  af-x:quantifies ?mass;
                  qudt:numeric value ?result;
                  qudt:unit uo:gram.
    ?mass         a pato:mass.
    ?loc          a envo:garden;
                  af-x:has value "FLORAL PARC"^^xsd:string.

    FILTER ( ?result > 20 && ?result < 100 && ?mtime >= "20181001"
        ^^xsd:date && ?mtime <= "20160101"^^xsd:date )
} LIMIT 100
```

Listing 1.2. SPARQL Query 2

```
Prefix: swrlb: <http://www.w3.org/2003/11/swrlb#>
Prefix: lcy: <http://vocab.org/lifecycle/schema#>
Prefix: af-m: <http://purl.allotrope.org/ontologies/material#>
Prefix: af-r: <http://purl.allotrope.org/ontologies/result#>
Prefix: af-rl: <http://purl.allotrope.org/ontologies/role#>
Prefix: af-p: <http://purl.allotrope.org/ontologies/process#>
Prefix: af-x: <http://purl.allotrope.org/ontologies/property#>
Prefix: qudt: <http://qudt.org/schema/qudt#>
Prefix: ro: <http://purl.obolibrary.org/obo/>
Prefix: envo: <http://purl.obolibrary.org/obo/>
Prefix: chebi: <http://purl.obolibrary.org/obo/CHEBI>

af-m:material(?sample), af-rl:sample role(?sr),
ro:has role(?sample, ?sr), envo:garden(?loc),
ro:located in(?sample, ?loc), af-p:test(?test),
ro:realized in(?sample,?test), af-r:scalar quantity datum(?data1),
af-x:has data output(?test, ?data1),
af-r:scalar quantity datum(?data2),
af-x:has data output(?test, ?data2),
chebi:mercury(?mercury), chebi:lead(?lead),
ro:bearer of(?test, ?mercury), ro:bearer of(?test, ?lead),
af-x:quantifies(?data1, ?mercury), af-x:quantifies(?data2, ?lead),
qudt:numeric value(?data1, result1),
qudt:numeric value(?data2, result2),
greterThan(?data1 > 15), swrlb:greterThan(?data2 > 22)
-> lcy:state(?loc, "Garden under surveillance")
```

Listing 1.3. SWRL Statement

In general, the domain expert has no background knowledge in linked data techniques and therefore it is difficult for him to express SPARQL queries or SWRL statements. Instead, visual approaches can be helpful in that they provide graphical support for query building [12,21]. We aim to use these approaches in the implementation phase of our methodology.

The use case we have just presented showed that it is possible to:

1. Extend our domain-independent ontology to support the requirements of a domain-specific LIMS, e.g., agronomy domain in the use case.
2. Specify business rules and answer queries that are relevant for the domain expert.
3. Perform (1) and (2), even for a LIMS that have not been built with semantic capabilities. In the case study, we demonstrated how that is possible using mappings between the relational model (used by the LIMS of AgiLab company) and the ontology obtained by extending SECO.

5   Related Work

Since the inception of linked data standards, many solutions [20, 22, 24] for representing and formalizing scientific experiments have been proposed.

Brahaj et al. [2] proposed the Core of Scientific Metadata Ontology (CSMO). It has been designed based on the Open Source e-Research Environment[13] (eSci-Doc) applications with a focus on building e-research repositories and a virtual research environment that can assist researchers during their day-to-day work.

ISA (Investigation/Study/Assay) [17] is a data annotation framework for managing a wide variety of life science, environmental, and biomedical experiences. González et al. [9] have created linkedISA, a software component that transforms the ISA-Tab format into RDF. linkedISA defines mappings that link data to ontologies such as the Ontology for Biomedical Investigation[14] (OBI) and other core ontologies belonging to the Open Biomedical Ontologies (OBO) Foundry.

Another approach, Allotrope Foundation Ontologies (AFO) has been developed by the Allotrope Foundation[15] (AF) consortium and provides a shared controlled vocabulary and semantic model for the annotation of laboratory-generated data. AFO extends the top-level ontology Basic Formal Ontology (BFO) and other domain ontologies aligned with BFO.

Other projects have focused on specific domains. For example, in the field of biology, OBO Foundry [19] has proposed an extensive list of ontologies specifically for experiments in the biomedical domain.

The OBO Foundry has defined numerous best practices[16] to reuse existing ontologies, which have been accepted by a large community. There are also interesting rules [10, 11, 18, 23] specified by other semantic web communities.

Based on these guidelines and the requirements of our approach, we concluded that the CSMO is very limited in terms of documentation and updates. The ontology is out of date and is not developed by a community. There were also some modeling errors, specifically, the persistence of entities (IDs). Furthermore, the entities of CSMO are not useful in the case of scientific data stored in a LIMS.

linkedISA does not represent an ontology but is a tool for converting the ISA-Tab format to the Resource Description Framework (RDF) format by defining mappings to domain ontologies. However, these ontologies will be different as each one depends heavily on a distinct ISA project. Therefore, linkedISA is only relevant in ISA-based projects.

The main advantage of AFO is that it is based on laboratories' needs, their scientific vocabulary and measurements workflows. The vocabulary of AFO can be easily extended to cover multiple fields as:

---

[13] eSciDoc : https://www.escidoc.org.

[14] OBI Ontology: http://obi-ontology.org.

[15] Allotrope Foundation : https://www.allotrope.org/about-us.

[16] OBO Foundry best practices: http://www.obofoundry.org/principles/fp-000-summary.html.

- It covers multiple application domains by reusing and specializing existing ontologies (AFO vocabulary is aligned with the BFO).
- It defines persistent entities that are annotated by domain experts.
- It is still in active development and being used by a large number of laboratories from different fields.

However, AFO is still limited as it specified only some laboratory experiments' case studies. We showed throughout this paper extensions that we added to the AFO vocabulary to create a domain-dependent ontology, and can be extended to meet the requirements of domain-specific LIMS. Furthermore, our work is in line with the FAIR initiative [25] in that we seek to better document, annotate and curate scientific experiments, thereby improving their accessibility and reuse (see Sect. 3 for more details).

6   Conclusion

The detection of the semantics of data generated by laboratories is a complicated task that involves several dimensions: scientific data annotation, domain-dependent knowledge, scientific experiments reproducibility, context modeling, etc.

In this paper, we presented a domain-independent methodology for modeling the ontology that captures the domain of interest to the scientist. We have presented a core-ontology allowing the representation of scientific experiments' data. The genericness of the proposed ontology grants its use by all LIMS-based laboratories by easily extending it to cover different fields. Moreover, we have provided a method of specifying and combining constraints and business rules. However, the defined constraints may have different meanings depending on the context [7] in which they are used. For example, the constraint of considering that two samples are similar is valid only if they have been analyzed within the same parameters and are taken from near or same locations.

In the case of a R&D laboratories, we define a scientific context as any information that can be used to characterize a scientific experiment. We distinguish between two types of information: local (analysis, parameters, measurements, etc.) and surrounding (time, space, etc.).

The incorporation of constraints and rules is still to be investigated as part of future work. We also aim to propose a generic model of scientific contexts making explicit the scientist's knowledge by combining these business rules. The proposed context model will be related to the SECO entities and allow comparison between them using defined semantic similarity functions.

References

1. Arp, R., Smith, B., Spear, A.D.: Building Ontologies with Basic Formal Ontology. MIT Press, Cambridge (2015)
2. Brahaj, A., Razum, M., Schwichtenberg, F.: Ontological formalization of scientific experiments based on core scientific metadata model. In: Zaphiris, P., Buchanan, G., Rasmussen, E., Loizides, F. (eds.) TPDL 2012. LNCS, vol. 7489, pp. 273–279. Springer, Heidelberg (2012). https://doi.org/10.1007/978-3-642-33290-6_29
3. Brinkman, R.R., et al.: Modeling biomedical experimental processes with OBI. J. Biomed. Semant. **1**, S7 (2010). BioMed Central
4. Ciccarese, P., Ocana, M., Castro, L.J.G., Das, S., Clark, T.: An open annotation ontology for science on web 3.0. J. Biomed. Semant. **2**, S4 (2011). BioMed Central
5. Das, S.: R2RML: RDB TO RDF mapping language (2011). http://www.w3.org/TR/r2rml/
6. Degtyarenko, K., et al.: ChEBI: a database and ontology for chemical entities of biological interest. Nucleic Acids Res. **36**(suppl-1), D344–D350 (2007)
7. Dey, A.K.: Understanding and using context. Pers. Ubiquit. Comput. **5**(1), 4–7 (2001)
8. Fritzsche, D., et al.: Ontology summit 2016 communique: ontologies within semantic interoperability ecosystems. Appl. Ontol. **12**(2), 91–111 (2017)
9. González-Beltrán, A., Maguire, E., Sansone, S.A., Rocca-Serra, P.: linkedISA: semantic representation of ISA-TAB experimental metadata. BMC Bioinf. **15**(14), S4 (2014)
10. Gruber, T.R.: Toward principles for the design of ontologies used for knowledge sharing? Int. J. Hum Comput Stud. **43**(5–6), 907–928 (1995)
11. Gyrard, A., Serrano, M., Atemezing, G.A.: Semantic web methodologies, best practices and ontology engineering applied to Internet of Things. In: 2015 IEEE 2nd World Forum on Internet of Things (WF-IoT), pp. 412–417. IEEE (2015)
12. Haag, F., Lohmann, S., Siek, S., Ertl, T.: QueryVOWL: a visual query notation for linked data. In: Gandon, F., Guéret, C., Villata, S., Breslin, J., Faron-Zucker, C., Zimmermann, A. (eds.) ESWC 2015. LNCS, vol. 9341, pp. 387–402. Springer, Cham (2015). https://doi.org/10.1007/978-3-319-25639-9_51
13. Harris, S., Seaborne, A., Prud'hommeaux, E.: SPARQL 1.1 query language. W3C Recomm. **21**(10), 778 (2013)
14. Horrocks, I., Patel-Schneider, P.F., Boley, H., Tabet, S., Grosof, B., Dean, M., et al.: SWRL: a semantic web rule language combining owl and RuleML. W3C Memb. Submiss. **21**, 79 (2004)
15. Lassila, O., Swick, R.R.: Resource description framework (RDF) model and syntax specification (1999)
16. Vazquez-Naya, J.M., et al.: Ontologies of drug discovery and design for neurology, cardiology and oncology. Curr. Pharm. Des. **16**(24), 2724–2736 (2010)
17. Rocca-Serra, P., et al.: ISA software suite: supporting standards-compliant experimental annotation and enabling curation at the community level. Bioinformatics **26**(18), 2354–2356 (2010)
18. Simperl, E.: Reusing ontologies on the semantic web: a feasibility study. Data Knowl. Eng. **68**(10), 905–925 (2009)
19. Smith, B., et al.: The OBO foundry: coordinated evolution of ontologies to support biomedical data integration. Nat. Biotechnol. **25**(11), 1251 (2007)
20. Soldatova, L.N., King, R.D.: An ontology of scientific experiments. J. R. Soc. Interface **3**(11), 795–803 (2006)

21. Soylu, A., Giese, M., Jimenez-Ruiz, E., Kharlamov, E., Zheleznyakov, D., Horrocks, I.: OptiqueVQS: towards an ontology-based visual query system for big data. In: Proceedings of the Fifth International Conference on Management of Emergent Digital EcoSystems, pp. 119–126. ACM (2013)
22. Stoeckert, C.J., Parkinson, H.: The MGED ontology: a framework for describing functional genomics experiments. Int. J. Genomics **4**(1), 127–132 (2003)
23. Vandenbussche, P.Y., Vatant, B.: Metadata recommendations for linked open data vocabularies. Version **1**, 2011–2012 (2011)
24. Visser, U., Abeyruwan, S., Vempati, U., Smith, R.P., Lemmon, V., Schürer, S.C.: Bioassay ontology (BAO): a semantic description of bioassays and high-throughput screening results. BMC Bioinf. **12**(1), 257 (2011)
25. Wilkinson, M.D., et al.: The fair guiding principles for scientific data management and stewardship. Sci. Data **3**, 160018 (2016)

Email Business Activities Extraction and Annotation

Diana Jlailaty[✉], Daniela Grigori, and Khalid Belhajjame

Paris Dauphine University, Paris, France
diana.jlailaty@gmail.com, daniela.grigori@dauphine.fr,
kbelhajj@googlemail.com

Abstract. Emails play, in the personal and particularly in the professional context, a central role in activity management. Emails can be harvested and re-engineered for understanding the undocumented business process activities and their corresponding metadata. Our goal in this paper is to recast emails into business activity centric resources. We describe an approach that is able to discover business process activities from emails. In addition, for each activity type, we extract metadata such as the roles of the people exchanging the email, type of the attached documents, or the domains of the mentioned links. In order to extract activities from emails, we compare several popular non-linear classification techniques. Activities are then clustered according to their types, which allows us to construct the metadata for each activity type. We validate our approach using a public email dataset.

Keywords: Email mining · Text mining · Business process · Business activity · Process instance

1 Introduction

A recent study has shown that email is still the primary method of communication, collaboration and information sharing[1]. Emails have evolved from a mere communication system to a mean of organizing complex activities (workflows), storing information and tracking activities. They are nowadays used for complex activities ranging from the organization of events, to sharing and editing documents, to coordinating the execution of tasks involving multiple individuals.

To facilitate personal email management, a number of research proposals have been made, see e.g., [5,7,22]. For example, Corston et al. [7] exploit emails to identify actions (tasks) in email messages that can be added to the user's "to-do" list. While useful, current emails management tools lack the ability to recast emails into *into business activity centric resources*. In particular, our work targets the group of people or analysts that would like to apply analytics on emails

---

[1] http://onlinegroups.net/blog/2014/03/06/use-email-for-collaboration/.

D. Kotzinos et al. (Eds.): ISIP 2018, CCIS 1040, pp. 69–86, 2019.
https://doi.org/10.1007/978-3-030-30284-9_5

upon extraction of business activities and their metadata. Thus, the elicitation of business activities from a set of emails opens up the door to activity analytics by leveraging emails to answer analytic queries such as:

Q_1 What are the business activities executed by a specific employee? (to identify time-consuming tasks that are not known to be assigned to him).

Q_2 How many times a user applied an activity? (for example, an employee may wish to know how many times he/she applied for a travel grant).

Q_3 What are the groups of people doing similar work? (similar activity types).

To allow for the extraction of business activities from emails, and therefore cater for the evaluation of analytic queries such as the one listed above, a previous work [15] was proposed with its associated framework [13]. The associated framework allows for the discovery of process models that connect activities using data and control dependencies using existing process mining solutions [21]. While the usefulness of such a proposal was empirically validated, it also underlined some limitations of the method used for extracting activities. In particular, each email is associated with a single business activity. However, in practice, an email can be associated with 0 or multiple activities. Moreover, to perform analytics on the activities that are extracted, annotations describing the activities need to be provided. In fact, the extraction and analysis of such information help answering some other queries such as:

Q_4 What kind of documents are sent as email attachments for a specific activity?

Q_5 What domains of web pages or links are used for a given activity?

Q_6 Who are the people involved in an activity (that executed the activity or were informed about it).

Discovering business activities in emails is considered as a step forward towards building the overall business processes described in the email log. Besides business activities discovery, the ability to distinguish between different process instances is crucial for deducing the business processes. Emails process instances discovery was studied in [14].

In addition to giving an insight on activity resources, this information could be used for making suggestions when editing a new email. Suggestions about attachments or links to be included would allow to improve the communication. For instance, suggestions to attach a receipt for confirming payment activity could reduce the number of email exchanges.

With the above applications in mind, we propose in this paper, a solution for extracting business activities from emails, and for annotating the elicited activities. Specifically, we make the following contributions:

- We present a solution for eliciting business activities from emails using sentence extraction and clustering techniques. We discover multiple activities for each email.
- We automatically associate each activity type with a set of metadata that describes it.

– We evaluate our approach on a set of data from a public email dataset (Enron email log).

To the best of our knowledge, there are no proposals in the literature that caters for the extraction of business activities and their metadata from a set of emails, where each email can be associated with multiple activities.

The paper is organized as follows. We start by providing a brief motivating scenario that shows the challenges we address in this work, followed by an overview of our approach in Sect. 2. Sections 3, 4, 5, and 6 explain our approach in details. Experiments and results are provided in Sect. 7. An analysis of the related work is provided in Sect. 8. The paper is finally concluded in Sect. 9.

2   Motivation and Approach Overview

It is widely recognized that emails have the potential of apprising the business activities that are conducted by individuals and companies. With this in mind, a number of proposals in the literature have aimed at recognizing activities by harvesting emails (see e.g., [7,17]). For example, in a previous work [15], it was shown that the body of the emails can be scanned to identify the business activity the email encompasses. While useful, this work, as well as existing proposals, assume that each email is associated with at most one activity. This assumption does not hold in practice, however. To illustrate this, consider the following email example taken from the Enron email log.

From: thomas.myers@enron.com
To: bus.all-hou@enron.com
Subject: Manual Wire and Same Day Payment Authorization

Dear all,
I am pleased to announce that it was an effective day (8/16). We authorized signing manual wires. Same day Tom Myers transfer payments to Georgeanne Hodges. Attached is the list of payments. As always Wes Colwell will also continue to have signing authority.
Thanks,
Tom

Manual examination of the above email reveals that it encompasses the following activities: *"Authorize signing manual wires"*, *"Transfer payments"*, *"Keep signing authority"*. Note, however, that using existing solutions, e.g. [15], will yield a single business activity, namely *"Transfer payments"* since the email is clustered with other emails concerned in the payment activity. However, the activity *"Transfer payments"* is one of multiple business activities mentioned in the email.

The above example shows the need for a solution that is able to elicit potentially multiple activities from a single email. We note also that emails are sometimes associated with resources, e.g., attachment files, web links, actors... For example, the above email is associated with an attachment file containing the list of payments. Such resources can be harvested to derive metadata that can be utilized to annotate elicited activities. The derived annotations can be utilized to help better understanding of the activities and to facilitate their management, e.g., their indexing and search.

Elaborating a solution that is able to elicit multiple activities from an email, and to augment such activities with annotations, raises a number of challenges that need to be addressed.

C_1 An email does not provide only information on business activities, it also provides information on non-business oriented activities. We, therefore, need to be able to distinguish between business and non-business activities (non-business oriented: *announce an effective day*; and business-oriented: *authorize signing manual wires, transfer payments, have signing authority*)

C_2 Not all emails contain information about business activities. We, therefore, need to be able to identify and discard non-relevant emails.

C_3 To be able to answer analytic queries $Q_1 - Q_3$ described in the introduction, we need to identify all the emails describing multiple occurrences of the same activity (e.g., all the emails about "transfer payments"). This is challenging, as emails are free texts and can use different words to describe the same activity.

C_4 An email may encompass multiple business activities, and as such associating the appropriate metadata with each activity is not obvious. We, therefore, need a means to correlate the information that can be extracted and abstracted from the resources associated with the emails (such as attached files, links and actors) with their corresponding business activities to answer queries like Q_4, Q_5, Q_6.

2.1   Approach Overview

We address the challenges we have just described using the approach depicted in Fig. 1. It takes as input an email log and produces the set of business activity types it contains, associated with their corresponding metadata.

The main phases of our approach are:

– **Phase 1: Data Preprocessing:** Since the important information of an email is contained in its body and subject unstructured texts, some preprocessing should be applied before any analysis. Texts are cleansed, segmented into separate sentences, and finally, verb-noun (verb-object) pairs are extracted for each sentence.

– **Phase 2: Relevant Sentences Extraction:** The goal of this phase is to extract from each email the sentences that contain business activities or information about activities (challenge C_1). For doing so, we use a machine learning based approach to classify sentences into two categories: process-oriented or not.

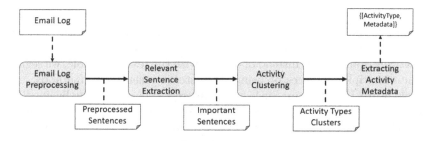

Fig. 1. Approach overall steps

As a consequence, emails that do not contain such sentences are considered to be non-business oriented emails and will be excluded from our analysis (challenge C_2). In order to decide whether the verb-noun pairs in the sentences are process oriented, we use domain specific knowledge (vocabularies extracted from process model repositories or ontologies).

- **Phase 3: Activity Clustering:** The goal of this phase is to identify all the emails containing occurrences of activities of the same type in order to answer analytical queries (challenge C_3). The input of this step is the subset of the extracted relevant sentences that contain business activities. Clustering techniques are applied to group together sentences describing activities of the same type. Cluster labels will be provided in a semi-automatic way in order to suggest activity names.

- **Phase 4: Extracting Activity Metadata:** In this phase, each activity type (represented by a cluster) will be associated with some metadata deduced from the set of information associated with the activity instances contained in the cluster. The metadata includes information like the role of the sender(s)/receiver(s) of the exchanged email, information about the resources that it uses (sent documents, information about the domain of web pages included in its description) and the actors performing the activity. This step addresses the challenge C_4.

3   Phase 1: Email Log Preprocessing

The input data of our approach is an email log. An email log is a set of emails exchanged between different entities (people, companies, etc.) for a specific purpose such as scheduling a meeting, organizing a conference, or purchasing an item etc. Each email is characterized by multiple attributes: subject, sender, receiver, body and timestamp. The most important information in an email is found in its subject and body texts, which are unstructured data-types. These texts should be translated into a format that is expected by the analysis tools. For this reason, data preprocessing is applied:

1. **Cleansing:** Unstructured data and particularly texts may contain redundant, missing or useless content. The cleansing step eliminates such kind of content that would have a negative effect on the analysis phase.
2. **Text Segmentation:** Instead of dealing with the text as a whole, we segment it into multiple sentences by using the sentence tokenization methods.
3. **Verb-noun Extraction:** Since we are interested in discovering business process activities we extract the verb-noun pairs that are likely to be candidates of such activities. We use for this the Stanford Parser which outputs grammatical relations in the new Universal Dependencies representation [8]. The following verb noun pairs will be extracted from the example email: *Sentence 1: Null; Sentence 2: pleased announce, announce day; Sentence 3: authorized signing, signing wires; Sentence 4: transfer payments; Sentence 5: attached payments; Sentence 6: continue signing, signing authority; Sentence 7: Null; Sentence 8: Null.*

4   Phase 2: Relevant Sentence Extraction

Not all the sentences that compose an email are about business activities. Some sentences may be talking about personal issues or greetings. We, therefore, need a means for identifying the sentences in the email that provide information about business activities such as the verb-noun pair representing the business activity or some other information characterizing the activity such as names of actors, locations, documents, links etc. We refer to such sentences by *relevant sentences*. To identify relevant sentences in an email, we use a classification technique that associates each email sentence with one of the following labels *Relevant* or *Non-Relevant*. This problem of sentence classification can be related to the problem known in the literature as extractive summarization [11], given that it reduces an input text by discarding non-informative sentences.

While machine learning classification techniques have been thoroughly investigated in the literature, it is widely recognized that there is no *one-size-fits-all* solution. Indeed, there are two parameters that considerably affect the outcome of classification: (i) The features used to characterize the population, in our case email sentences, and (ii) the classification algorithm used. In the remaining of this section, we discuss these parameters. In doing so, we present arguments that justify our choices.

4.1   Feature Extraction

We define the following features that characterize the sentences as relevant and business oriented. We divide the features into two categories: *Syntactic Features* and *Semantic Features*.

1. **Syntactic Features:**
   - F_1: **Position of a sentence:** This feature provides the position of the sentence in an email. Usually, the most important sentences in an email

are the ones that are located in the middle. The sentences located in the beginning and the end of an email are more likely to be introductory or ending phrases with no significant amount of information.

- F_2: **Length of the sentence:** Usually short sentences are not as important as long ones [1]. We assume that as the number of words increases in a sentence, the amount of information it holds increases too.

2. **Semantic Features:**
    - F_3: **Number of named entities:** The existence of named entities such as names of people (actors), locations, names of conferences, etc. gives an indication about the possibility that this sentence contains an activity (which is performed by the mentioned actors or in the specified location).
    - F_4: **Cohesion of a sentence with the centroid sentence vector:** We define the centroid sentence as the entity holding the overall information about an email. Email sentences with higher cohesion (higher similarity value) with the centroid sentence are considered of a higher importance. Using the word embeddings techniques, each email sentence is converted into a numerical vector (by averaging the numerical vectors of its words). The centroid sentence vector is deduced by calculating the average of all the numerical vectors of the email sentences.
    - F_5: **Dissimilarity with greetings and ending phrases:** The greeting and ending phrases are considered to be of no importance for our analysis. Thus email sentences with higher similarity with these phrases are likely to be less important in our analysis. The dissimilarity is computed based on the distance between each sentence and an already defined set of email greetings and ending phrases.
    - F_6: **Similarity with the process-oriented activities:** This feature estimates the likelihood of the sentence to contain a process-oriented activity. A business process is a set of business activities executed towards the achievement of some business goals. A business activity is a step of the business process which performs a part of the overall process goal. Usually, activities in the emails are written in the form of verb-object pairs such as "confirm meeting", "cancel meeting", "send payment", "receive receipt", etc. So we estimate the probability that the verb-noun pair of a sentence is, in fact, a business activity.
    We calculate it based on the semantic similarity between the verb-noun pairs in a sentence and a collection of process-oriented activities. This collection can be built by extracting activities from different repositories of process models including different topics such as purchasing or selling items, insurance agreements, incident management, money transfer, meeting organization, etc.

It is important to note that all word and sentence semantic similarity calculations applied in this work are done using Word2vec model [10]. Word2vec is a computationally-efficient method for learning high-quality distributed vector representations (called word embeddings). More precisely, Word2vec model characterizes each word by a numerical vector. By averaging the numerical vectors of the words of a sentence, we obtain the sentence's feature vector.

What is missing in our training data is the label or class of each sentence. We label our training data manually. An expert decides whether the sentence is meaningful from a business-oriented perspective (it contains a business activity or any information about an activity) or not. By having our training data built, we are ready now to train our model.

4.2   Classification

In order to ensure we obtain the most efficient classification results, we compared 7 different *non-linear* classification techniques: (1) Neural Networks (NN) (2) Non Linear Discriminant Analysis (NLDA) (3) K-Nearest Neighbor (KNN) (4) Decision Tree (DT) (5) Gradient Boosting Classifier (GBC) (6) Gaussian Naive Bayes (GNB) (7) Support Vector Machine (SVM). In the experimentation Sect. 7, we explain how we deduce that the Gradient Boosting classification technique is the most efficient method in our case. The trained model is used to classify the sentences of the incoming emails as important or not from a process-oriented perspective. For each sentence, its feature vector is obtained and fed to the trained model which decides whether the sentence is relevant or not.

Going back to our example, the relevant sentences are the following: ***Sentence 3***: We authorized signing manual wires, ***Sentence 4***: Same day Tom Myers transfer payments Georgeanne Hodges, ***Sentence 5***: Attached list payments, ***Sentence 6***: As always Wes Colwell also continue have signing authority. We can see that the extracted sentences are either the sentences containing the business activities (sentences 3, 4, and 6) or sentences containing annotations about activities (sentence 5).

5   Phase 3: Activity Types Discovery

The business activities elicited in the previous steps can be further processed to organize them by their *activity type*. Organizing business activities by types has several applications, e.g., they can be used for indexing activities, searching for them and performing analytic queries over them. For example, a company manager may be interested in knowing the average number of business activities of type "shipping goods" that are processed by month.

To organize activities by type, we use clustering techniques. Since we do not have a priori knowledge about the number of activity types, hierarchical clustering is chosen as it does not require predefining this constraint. Hierarchical clustering is applied to sentences containing process oriented verb-noun pairs (i.e. activities). The similarity between two sentences is calculated using the cosine similarity between the Word2vec vectors of their verb-noun pairs (i.e. activities). For each verb-noun pair, an average numerical feature vector is obtained. The distance function used for clustering is:

$$Sim(a_i, a_j) = CosineSim(vn_i, vn_j) \tag{1}$$

where vn_i and vn_j are the verb-noun pairs of the activities

According to our visualization study, we apply a cut on the obtained cluster hierarchy. We try multiple cuts on multiple levels to deduce that which provides the best clustering quality. This phase gives as a result a set of clusters where each cluster contains sentences from different emails but with the same activity type ($\{AC_i\}$).

Activity Labeling. After clustering activities contained in emails, our goal now is to deduce their labels. In other words, we provide labels for the obtained clusters $\{AC_i\}$. For each cluster, we choose the top N verb-noun pairs mentioned in the sentences (for example N can be equal to 3). Then one of these verb-noun candidates is chosen by an expert as a label for the cluster and consequently the activity type.

6 Phase 4: Extracting Activity Metadata

The aim in this phase is to enrich the activity types with metadata that provides the user with further analytical capabilities. For example, a manager may want to know what are roles of the employees that participate in an activity type. Another application would be suggesting for a user the document type that should be attached to an email containing a specific activity type. Using the classified relevant sentences obtained in the previous phase, we apply the two following steps: (A) extracting information related to each activity type instance in each cluster AC_i and (B) aggregating the extracted information of these activity instances to formulate the metadata for the whole cluster i.e. for the activity type represented by the cluster.

We define metadata schema that can be applied on any activity type (independent from an activity type on business domain). However, different values of these attributes are associated to different activity types. So the metadata schema is domain independent, but the values provided are dependent.

6.1 Extracting Information of Each Activity Type Instance

We extract/infer for each email activity (cluster of activity instances) the following attributes when applicable:

– **Organizational role of the people exchanging the email:** We make here the *reasonable* assumption that different occurrences (instances) of the same activity type are exchanged between people of similar organizational roles. The roles of people are deduced from the organizational database. For example, in a company, each email address of an employee will be associated with his/her role in a predefined knowledge-base or directory such as administrator, engineer, etc.We use a database which contains for each email address its corresponding role such as *administrator, director*, etc.
– **Actors:** The sender(s)/receiver(s) of an email containing an activity are people who may be interested to know information about the activity (involved

in the activity). However, the actors are the people that are actually performing the activity described in the email. Actors can be inferred by checking the names of people mentioned in the body of the email. The actors are extracted by distinguishing the person names from the named entities in the sentences.

- **Attachments types:** We characterize the attached documents such as a bill, document to be filled, filled document, data document, etc.
- **Web pages descriptions:** such as web-pages and their domains and descriptions.

To find the values for the above attributes, we apply the following steps:

Step 1 Since in each email we may have multiple activities, we need to correlate each activity with the email sentence(s) containing information related to this activity. This is done by using:

- The semantic similarity between the words of the verb-noun pair (the activity) and the words of the other relevant sentences. Sentences including information about an activity are likely to contain words similar to the activity name. For example, suppose we have an email containing the following activity sentence *"We received your payment"* and information sentence *"The receipt of your payment is attached to this email"*. Both sentences contain words about *"receiving"* and *"payment"*. Using the semantic similarity measurements, we can detect the correlation between both sentences.
- The farness between the activity sentence and information sentence. We suppose that correlated sentences are likely to be close to each other.

Step 2 For each activity, we parse the sentence containing the activity instance and its correlated information sentences to extract the values of the attributes.

6.2 Eliciting Activity Type Metadata

After extracting information for each activity type instance in the cluster, we aim to aggregate this information to get a generalized definition of the metadata describing the corresponding activity type.

For instance, we check the attachment type correlated with all the activity instances of specific type and deduce the most occurring one to be associated with the activity type. Regarding the weblinks, we concatenate keywords describing the weblinks correlated with the activity instances of the same activity type. The set of concatenated keywords is associated with the corresponding activity type as a weblink description.

For the sender(s)/receiver(s) roles and actors attributes, we keep the information as they are without any aggregation.

7   Experimentation and Results

In this section, we describe in details the experiments we did to validate our approach.

7.1 Experimentation Settings

Dataset. In order to do our experimentation, we use the Enron dataset [16] which contains several folders belonging to multiple employees from the enterprise Enron. Each email from the log loaded from a chosen folder can be divided into a set of attributes: sender, receiver, subject, body, and timestamp. For the training phase, we choose an email folder containing 628 emails, which include in total 5124 sentences.

Implementation Packages and Tools. We used python packages to implement the described approach. The preprocessing step is applied using the Natural Language Toolkit (NLTK)[2] which is an open source Python library for Natural Language Processing. In the second phase, we extract the features values for each sentence. Specifically, for the extraction of named entities, we use the tokenization and chunking functions of NLTK. For calculating semantic similarities, we use the Word2vec model. We import the Gensim python package in which we load a 3.4 GB Word2vec model containing all vectors of 1 billion words trained on Google news corpus. For computing the similarity between verb-noun pairs of the email sentences and the process-oriented activities, we build the process-oriented activities set. This is done by obtaining an external process model repository[3] containing about 4000 models of different domains. Using this repository we extracted about 21000 business process activities. The extracted verb-noun pairs (using Stanford parser[4]) from the email sentences are then compared to the activities contained in the built set.

In order to build the training data set, we manually assign labels for each sentence. The labels refer to whether the sentence should be included or excluded from the analysis. The expert chooses 1 as a label if the sentence contains business process oriented activities information, 0 otherwise.

We then train 7 different classification models: Non-linear discriminant analysis, neural network, k-nearest neighbor, decision tree, gradient boosting classifier, Gaussian naive Bayes, and support vector machine. The most efficient trained model is used later for the classification of sentences. All training algorithms are loaded in python using the Sklearn package.

7.2 Classification

As described in Sect. 4, classification is applied to distinguish between business and non business oriented sentences. Using the already labeled data, we test the reliability of the 7 classifiers by applying the k-cross validation technique. We choose k = 5. Table 1 shows the results of the reliability test for the seven classifiers proving that GBC is the most reliable.

---

[2] https://www.nltk.org/.

[3] https://www.signavio.com/fr/products/process-manager/.

[4] https://nlp.stanford.edu/software/lex-parser.shtml.

Table 1. k-cross validation reliability measurement

Evaluator	NN	NLDA	KNN	DT	GBC	GNB	SVM
k = 1	0.5806	0.6193	0.5698	0.6129	0.6236	0.5913	0.6144
k = 2	0.6413	0.6717	0.6521	0.6282	0.72	0.6956	0.71
k = 3	0.5373	0.6153	0.5714	0.5714	0.5824	0.5714	0.55934
k = 4	0.6153	0.6373	0.5934	0.6812	0.6943	0.6362	0.6043
k = 5	0.6373	0.6123	0.6043	0.6702	0.7362	0.7322	0.7032

After training the classifiers, we choose a subset of Enron dataset containing 50 emails with 724 sentences to test the performance of the obtained classifiers. The sentences are transformed into the defined features to be used as an input for the trained models. To check which classifier has the best prediction efficiency, we measure the prediction efficiency using classifiers evaluators:

1. $Recall = \frac{TP}{TP+FN}$
2. $Accuracy = \frac{TP+TN}{TP+TN+FP+FN}$
3. $Precision = \frac{TP}{TP+FP}$
4. $F - measure = 2\frac{precision*recall}{precision+recall}$

where TP = True Positive, TN = True Negative, FP = False Positive, FN = False Negative, N is the total population. Table 2 below show the different classification evaluators values for the 7 different classifiers.

Using the results, we deduce that the Gradient Boosting Classifier (GBC) provides the best values which means it is the most efficient classification model for our goal. The sentences classified by the GBC are going to be clustered in the next step.

Table 2. Binary classifier performance evaluation

Evaluator	NN	NLDA	KNN	DT	GBC	GNB	SVM
Recall	0.68147	0.67934	0.63393	0.78164	0.78408	0.67099	0.67320
Accuracy	0.69952	0.70316	0.66134	0.78892	0.80333	0.70306	0.69760
Precision	0.69041	0.70217	0.65639	0.78661	0.79098	0.67793	0.67101
F-measure	0.71351	0.73246	0.67773	0.77543	0.81312	0.74402	0.74390

7.3 Clustering

In order to efficiently measure the performance of the clustering, we choose only the sentences that are correctly positively classified (True Positives). We use the Scipy[5] package to apply the hierarchical clustering on the activities of

---

[5] https://www.scipy.org/.

the positively classified sentences. We obtained 7 clusters where each cluster represents an activity type. The obtained activity types are: "Fill Document", "Modify Data", "Suggest Meeting", "Open Discussion", "Modify Rule", "Modify Capacity", and "Provide Product".

In a previous work [15], a multi-level clustering is applied to obtain email topics (first level) and activity types (second level). It was supposed that each email contains only one activity type. Each email is characterized by a numerical feature vector computed as the average of the Word2vec feature vectors of all its words. These feature vectors are then used to calculate the similarity between emails. Applying this approach as a baseline, only two clusters of the activities "Modify Data" and "Suggest Meeting" are obtained.

However, by examining 200 emails from a folder in Enron dataset, we discovered that 2% of the emails contain only one activity. In the current work, we conduct a finer granularity analysis. Instead of examining the email as a whole, we analyze email sentences.

Table 3 below shows the different clustering evaluation metrics values. Besides precision and recall (defined in the previous subsection), we use:

1. $Purity(\Omega, \mathbb{C}) = \frac{1}{N} \sum_k max |\omega_k \cap c_j|$ where $\Omega = \{\omega_1, \omega_2, \ldots, \omega_K\}$ is the set of clusters, $\mathbb{C} = \{c_1, c_2, \ldots, c_J\}$ is the set of classes and N is the total number of instances.
2. $Rand - Index = \frac{TP+TN}{TP+TN+FP+FN}$

Reaching this step, the questions Q_1 to Q_3 can be answered.

Table 3. Clustering performance evaluation

Clustering metrics	Current approach
Purity	0.86
Rand-Index	0.81
Precision	0.918
Recall	0.87

7.4 Metadata Extraction

The main information we extract for each activity are:

– **Organizational role of the people exchanging the email:** The role of sender(s)/receiver(s) could be obtained from the organizational database or directory of the enterprise. As this informations is missing in the Enron dataset, we have built a directory for the email addresses contained in the testing emails. We associate to each email address (of the people exchanging an email) a specific role such as administrator, director, engineer, etc.

- **Actors:** Each activity instance is associated with a set of relevant sentences. These relevant sentences (in addition to the sentence containing the activity) are analyzed in order to extract Named Entities of type "Person" using NLTK python packages. The extracted named entities are supposed to represent the actors of an activity.
- **Attachments types:** We build (simulate) a thesaurus containing names and extensions of attachments associated with their subjects or types. For example, the activity instances of type "Modify Data" are associated with attachments of names "Template.xls" and "Doc.xls". Thus we add these names to the thesaurus associated with document type as "excel data document to be edited". This thesaurus is always enriched and used to extract types for similar attachment names.
- **Web pages descriptions:** We use web scraping[6] in python to automate keywords extraction of each website mentioned in the email sentences. This tool loads the HTML file of a website. We extract words from website's texts that are most similar to the website title words.

  Table 4 shows an example on activity instances belonging to the same activity type "Modify Data" with their corresponding information.

Table 4. Activity instances of the same activity cluster (same activity type) and their associated information

Activity name	Sender(s)/Receiver(s) role	Attachment	URL description	Actors information
Modify data	Administrator	Template.xls	Global innovation economy	No actor
Edit data	Administrator	Template.xls	Insights on economy	Jennifer Stewart: Administrator
Add information	Engineer	Doc.xls	No description	No actor

The set of values of the metadata attributes can be inspected by the user. After applying data aggregation, we obtain for the activity type "Modify Data" the following information as metadata:

- Sender(s)/Receiver(s) Role: Administration, Engineering
- Attachment Type: excel data document to be edited.
- Web-page description: Global innovation and insights about Economy.
- Actor: Jennifer Stewart, role: Administrator

Having these values in hands, we can answer the question Q_4 to Q_6.

7.5 Discussing the Analytical Questions

Each activity type is now correlated with a set of attributes which facilitates answering question Q_1 to Q_6. For example, the answers for questions Q_2 and Q_6 are the following

---

[6] https://data-lessons.github.io/library-webscraping/04-lxml/.

Q_1 **What are the business activities executed by a specific employee? (to identify time-consuming tasks that are not known to be assigned to him).**
Each activity type is correlated to a set of actors which allows us to specify all the business activities an employee applies. For example, the employee Jennifer Stewart is responsible for the execution of the activity "Edit Data".

Q_2 **How many times a user applied an activity? (for example, an employee may wish to know how many times he/she applied for a travel grant).**
To answer Q_2, we count the number of activity instances of the same activity type applied by the same actor. For example, the activity "Edit Data" is applied 7 times by the employee Jennifer Stewart.

Q_3 **What are the groups of people doing similar work? (they apply similar activity types).**
People involved in an activity are the people exchanging emails about this activity which can be deduced from the sender/receivers part of the email. For example, the group of people who are involved in the activity "Modify Data" are: Stephen Allen, Tony B, Herb Caballero, Kenneth, Roger Raney, Henry Van, Linda Adels, Paul Duplachan

Q_4 **What kind of documents are sent as email attachments for a specific activity?**
Each activity type is characterized by an attribute describing the attachments usually associated with it. Since we didn't have the contents of the attachments of Enron dataset, we only can extract information from the name and the type of the attachments of a specific activity.

Q_5 **What domains of web pages or links are used for a given activity?**
Using the obtained keywords describing the links and web pages associated with activities, a user can deduce their domains.

Q_6 **Who are the people involved in a specific activity?**
To answer Q_6, we examine the set of actors associated with an activity type. These information can be presented as a graph where the relations between actors are represented by edges. For example, the actor Jennifer S. and Carrie R. are related (have similar tasks) if they both apply the same activity such as "Edit Data".

8   Related Work

We categorize the related works into the following categories:

Summarization of Emails. The volume of the received emails entails a great cost in terms of the time required to read, sort and archive the incoming data. Email summarization is a promising way to reducing this email triage. We focus here on abstractive summarization: extracting sentences from emails that are considered important for our analytical goals. The work of Alam et al. [18] uses parsing methods to abstract some keywords for each sentence in an email. The abstracted keywords are matched with the user's priority keywords to extract important sentences from emails. The work of Rambow et al. [19] focuses on creating a set of sentence features by using proven text extraction features (e.g., the relative position of the sentence in the document) and adding email-specific features (e.g., the similarity of the sentence with the email Subject field). Their results show that email-specific features significantly improved summarization.

Previous email summarization researches work on extracting important sentences from emails. However, in our work we extract only important sentences from a business process oriented point of view. This has two advantages: (1) helping the user to distinguish between business emails and personal emails (those that do not include any business-oriented sentence), (2) helping the user to better manage his/her emails by efficiently specifying the sentences that contain business activities and their information.

Email Organization. The common objective of the related works presented in this paragraph is to categorize emails into a set of classes (folders, topics, importance, main activities). In the work of Alsmadi et al. [2], a large set of emails is used for the purpose of folder classifications. Five classes are proposed to label the nature of emails users may have: Personal, Job, Profession, Friendship, and Others. Another work by Yoo et al. [23] develop a personalized email prioritization method using a supervised classification framework. The goal is to model personal priorities over email messages, and to predict importance levels for new messages using standard Support Vector Machines (SVMs) as classifiers.

In a previous work of Jlailaty et al. [15], a preliminary method is defined to discover and label activities from email logs based on hierarchical clustering. The distance function used for the clustering takes into account all the emails in the input folder and all the sentences of these emails. The similarity measures were experimentally compared based on word2vec and other traditional methods and it was concluded that word2vec based measures give better results. The work of Bekkerman et al. [3] represents emails as bag-of words (vectors of word counts) to classify them into a predefined set of classes (folders).

In our work, a finer granularity analysis is applied (on the level of sentences and not emails as a whole) allowing us to extract more information from an email, which is closer to the fact that people may exchange one email describing several tasks.

Extracting Tasks from Emails. In the work of Faulring et al. [9], tasks contained within sentences of emails are classified using 8 predefined set of classes of tasks.

In the same context, Cohen et al. [6] use text classification methods to detect "email speech acts". Based on the ideas from Speech Act Theory [20], the authorsdefine a set of "email acts" (e.g., Request, Deliver, Propose, Commit) and then classified emails as containing or not a specific act. Another work by Corston-Oliver et al. [7] identifies action items (tasks) in email messages that can be added by the user to his/her "to-do" list. Syntactic and semantic features are used to define the sentences of the emails to be classified using SVM as containing tasks or not.

In our work, we overcome the limitation of specifying a predefined set of activity classes. Our approach is able to cope with the diversity of activity types that can exist in emails. Moreover, we extract also activity attributes and metadata.

Email Task Management. Several applications mainly consider the process of associating manually the email and their metadata such as attachments, links, and actors with activities. TaskMaster [4] is a system which recasts emails into Thrasks (thread + task), the interdependent tasks which comprise threads. It is a method to organize the threads into tasks. In Gwizdka et al. [12], the TaskView interface is proposed for improving the effectiveness and efficiency of task information retrieval. They manage emails according to a timeline to facilitate the management of pending tasks embedded in messages. They assume that one message corresponds to one task.

In our work, we do not only work with low-level information (associating the attachments or links with tasks). We deduce high-level metadata such as attachments types and subjects, weblinks domains, actors, or people interested in an activity type. This will provide the user with better email analysis tools.

9   Conclusion

We presented in this paper an approach which takes as input an email log and extracts business activity types and their associated metadata. We applied our approach on Enron data and we provided an activity-centric analysis of emails. We are currently investigating how the extracted information and the classification models that we have built could be used for recommendations for incoming emails or editing emails: tasks to be added to the task list, annotations to facilitate email retrieval, suggesting attachments or persons to be added in the recipients' list, etc. Another future direction concerns the use of the proposed annotations for improving activity-oriented email management, e.g., their indexing and search.

References

1. Allahyari, M., et al.: Text summarization techniques: a brief survey. arXiv preprint arXiv:1707.02268 (2017)
2. Alsmadi, I., Alhami, I.: Clustering and classification of email contents. J. King Saud Univ.-Comput. Inf. Sci. **27**(1), 46–57 (2015)
3. Bekkerman, R.: Automatic categorization of email into folders: benchmark experiments on Enron and SRI Corpora (2004)
4. Bellotti, V., Ducheneaut, N., Howard, M., Smith, I.: Taskmaster: recasting email as task management. PARC, CSCW **2** (2002)
5. Bellotti, V., Ducheneaut, N., Howard, M., Smith, I., Grinter, R.E.: Quality versus quantity: e-mail-centric task management and its relation with overload. Hum.-Comput. Inter. **20**(1), 89–138 (2005)

6. Cohen, W.W., Carvalho, V.R., Mitchell, T.M.: Learning to classify email into speech acts. In: Proceedings of Empirical Methods in Natural Language Processing (2004)
7. Corston-oliver, S., Ringger, E., Gamon, M., Campbell, R.: Task-focused summarization of email. In: Proceedings of the Text Summarization Branches Out ACL Workshop (2004)
8. De Marneffe, M.-C., Manning, C.D.: Stanford typed dependencies manual. Technical report, Stanford University (2008)
9. Faulring, A., et al.: Agent-assisted task management that reduces email overload. In: Proceedings of the 15th International Conference on Intelligent User Interfaces, pp. 61–70. ACM (2010)
10. Goldberg, Y., Levy, O.: word2vec explained: deriving Mikolov et al'.s negative-sampling word-embedding method. arXiv preprint arXiv:1402.3722 (2014)
11. Gupta, V., Lehal, G.S.: A survey of text summarization extractive techniques. J. Emerg. Technol. Web Intell. 2(3), 258–268 (2010)
12. Gwizdka, J.: Taskview: design and evaluation of a task-based email interface. In: Proceedings of the 2002 Conference of the Centre for Advanced Studies on Collaborative Research, p. 4. IBM Press (2002)
13. Jlailaty, D., Grigori, D., Belhajjame, K.: A framework for mining process models from emails logs. arXiv preprint arXiv:1609.06127 (2016)
14. Jlailaty, D., Grigori, D., Belhajjame, K.: Business process instances discovery from email logs. In: 2017 IEEE International Conference on Services Computing (SCC), pp. 19–26. IEEE (2017)
15. Jlailaty, D., Grigori, D., Belhajjame, K.: Mining business process activities from email logs. In: 2017 IEEE International Conference on Cognitive Computing (ICCC), pp. 112–119. IEEE (2017)
16. Klimt, B., Yang, Y.: The enron corpus: a new dataset for email classification research. In: Boulicaut, J.-F., Esposito, F., Giannotti, F., Pedreschi, D. (eds.) ECML 2004. LNCS (LNAI), vol. 3201, pp. 217–226. Springer, Heidelberg (2004). https://doi.org/10.1007/978-3-540-30115-8_22
17. Kushmerick, N., Lau, T., Dredze, M., Khoussainov, R.: Activity-centric email: a machine learning approach. In: Proceedings of the National Conference on Artificial Intelligence, vol. 21, p. 1634. AAAI Press/MIT Press, Menlo Park/Cambridge (1999, 2006)
18. Kakkar, M., Alam, M.: Email summarization-extracting main content from the mail. Int. J. Innov. Res. Comput. Commun. Eng. 3 (2015)
19. Rambow, O., Shrestha, L., Chen, J., Lauridsen, C.: Summarizing email threads. In: Proceedings of HLT-NAACL 2004: Short Papers, pp. 105–108. Association for Computational Linguistics (2004)
20. Searle, J.R.: A classification of illocutionary acts. Lang. Soc. 5(01), 1–23 (1976)
21. van Dongen, B.F., de Medeiros, A.K.A., Verbeek, H.M.W., Weijters, A.J.M.M., van der Aalst, W.M.P.: The ProM framework: a new era in process mining tool support. In: Ciardo, G., Darondeau, P. (eds.) ICATPN 2005. LNCS, vol. 3536, pp. 444–454. Springer, Heidelberg (2005). https://doi.org/10.1007/11494744_25
22. Whittaker, S., Bellotti, V., Gwizdka, J.: Problems and possibilities. Commun. ACM Email PIM (2007)
23. Yoo, S., Yang, Y., Lin, F., Moon, I.-C.: Mining social networks for personalized email prioritization. In: Proceedings of the 15th ACM SIGKDD International Conference on Knowledge Discovery and Data Mining, pp. 967–976. ACM (2009)

Advanced Data Mining Techniques

Effective Pre-processing of Genetic Programming for Solving Symbolic Regression in Equation Extraction

Issei Koga[1] and Kenji Ono[2]([⊠])

[1] Graduate School and Faculty of Information Science and Electrical Engineering,
Kyushu University, 744 Motooka, Nishi-ku, Fukuoka, Japan
2IE17031W@s.kyushu-u.ac.jp
[2] Research Institute for Information Technology, Kyushu University,
744 Motooka, Nishi-ku, Fukuoka, Japan
keno@cc.kyushu-u.ac.jp

Abstract. Estimating a form of equation that explains data is very useful to understand various physical, chemical, social, and biological phenomena. One effective approach for finding the form of an equation is to solve the symbolic regression problem using genetic programming (GP). However, this approach requires a long computation time because of the explosion of the number of combinations of candidate functions that are used as elements to construct equations. In the present paper, a novel method to effectively eliminate unnecessary functions from an initial set of functions using a deep neural network was proposed to reduce the number of computations of GP. Moreover, a method was proposed to improve the accuracy of the classification using eigenvalues when classifying whether functions are required for symbolic regression. Experiment results showed that the proposed method can successfully classify functions with over 90% of the data created in the present study.

Keywords: Deep neural network · Pre-processing ·
Eigenvalue analysis · Genetic programming

1 Introduction

Estimating a form of equation that explains data is very useful to understand various physical, chemical, social, and biological phenomena. For instance, air and water, which are relatively simple operating fluids, are classified as Newtonian fluids, and their characteristics are determined by Stokes' law, which is introduced as an auxiliary equation (constitutive equation), and provide reasonable modeling of flow phenomena. However, constitutive equations for complex flow phenomena, such as non-Newtonian flows, multi-phase flows, and viscoelastic fluids, have not been obvious. If the form of the governing equations can be clarified, we can obtain a deeper understanding of the flow phenomena. A system of partial differential equations with a high degree of freedom is derived

© Springer Nature Switzerland AG 2019
D. Kotzinos et al. (Eds.): ISIP 2018, CCIS 1040, pp. 89–103, 2019.
https://doi.org/10.1007/978-3-030-30284-9_6

from the biological phenomena of material transfer in vivo. Constructing such a complicated system without domain knowledge is usually difficult. Therefore, estimating the form of the equation that expresses data is very useful in various fields of science, medicine, social science, and engineering.

Since measurement technology has also been developed with the progress of the development of sensors and analysis algorithms, many types of in-depth data can be obtained. Finding the relationship between variables in such measured data by the symbolic regression approach to express the nature of various phenomena would be hugely beneficial. However, the time cost required to find this relationship increases because the amount of data to be processed increases. Thus, finding a method by which to efficiently analyze a large amount of data is strongly expected. Genetic programming (GP) is often used to solve symbolic regression problems, i.e., modeling of data.

Schmidt et al. reported that the equations of motion and the Lagrangian can be re-discovered from observed data [1]. Therefore, GP has the ability to discover the governing equations of various phenomena from given data. They used GP to synthesize an equation combining various terms, which are generated from a given set of functions, i.e., candidate functions. In the case of complicated various phenomena, however, many types of candidate functions, e.g., trigonometric, exponential, and logarithmic functions, may be needed in order to construct the terms. Therefore, the computation time to find a solution increases exponentially because many variables and functions must be combined, resulting in a huge number of computations. In order to reduce the computing time, thus far, efficient search in a solution space should be performed by devising GP algorithms that include a fitness function and crossover. In addition, parallelization and GPU acceleration can improve the throughput of GP.

In the present paper, a novel method to reduce the computational complexity of GP is proposed. An effective candidate selection method using a deep neural network (DNN) is introduced and experiments and verification are performed.

The remainder of the present paper is organized as follows. Section 2 presents an overview of related research. Section 3 outlines the proposed method, and Sect. 4 presents details of the proposed method of reducing the computational complexity. In Sect. 5, experimental methods and the obtained results are discussed. Finally, Sect. 6 summarizes the present study.

2   Related Research

There are several methods by which to estimate expressions of formulas. Schmidt et al. proposed a method to simultaneously search both the form and the parameters of the formula representing the relationship between different physical quantities [1]. Their method successfully discovers the form and nature of a rational expression for a known physical system. Similar approaches have been used to construct models of data [2,4–7,9]. Martius et al. applied the candidate function to the activation function, and then created the structure of the expression from the internal structure of a neural network after learning [3]. However, functions for which the domain of definition is restricted, such as logarithmic functions, for example, cannot always be used as activation functions. In other words, since logarithmic functions cannot be used for the structure of the expression, data to which the method of Martius et al. can be applied are limited [3]. Although GP has the ability to synthesis various functions, its computational cost tends to increase as the number of combinations of candidate functions increases. Several improvements can be found in the literature to reduce the number of computations and improve the efficiency of calculation. The approximation or reasoning of a fitness function brings about effects such as the reduction of the number of computations and the prevention of local optimization [1,2,4–6]. In addition to the fitness function, crossover and extension of GP enable the symbolic regression problem to be solved efficiently [7–9]. As far as we know, however, no effective method that can be applied to multivariate complex data has been proposed. Thus, a novel method of reducing the number of candidate functions to be used for GP is proposed as a pre-processing because reducing the number of candidates can reduce the total computational cost of GP and improve throughput.

3   Reduction of the Number of Computations

In this section, how to reduce the number of candidate functions is described.

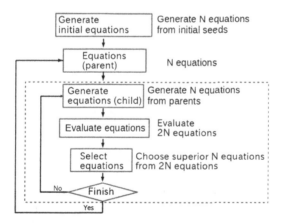

Fig. 1. Flowchart of a typical genetic programming process

The entire process of GP for the symbolic regression is shown in Fig. 1. The process of generating initial equations generates an initial set of functions using all possible candidate functions, which may include unnecessary functions for modeling. Therefore, the number of trials in GP using any algorithm increases exponentially due to the number of candidate functions. If the number of candidate functions can be reduced beforehand for GP, the number of computations for the GP process can be greatly reduced. This requires classifying all candidate functions as necessary or unnecessary candidate functions. This classification is required for the GP process to succeed. An evaluation function is created in order to determine whether each candidate function is necessary in modeling, and the candidate functions are then classified by a given threshold value. The evaluation function was generated by machine learning using a deep neural network. The details of the proposed method are described in the next section.

4   Proposed Method

In this section, an overview of a deep neural network and a method by which to reduce the computational complexity of GP are described.

4.1   Evaluation Function

A deep neural network is used to create the evaluation function that determines whether the candidate function is necessary for modeling. The output value of the evaluation function is in the range of $[0, 1]$, and if the value exceeds a given threshold value, the candidate function is judged to be necessary for GP. The structure of the neural network is shown in Fig. 2.

If the evaluation value (EV) is equal to or larger than a threshold value (TV), then the function becomes a candidate function and is passed to the next GP step, otherwise, the function is rejected. In this approach, since the neural network is used as a black box, a problem arises in that there are no reasonable grounds for the given TV. Thus, classification can be improved by introducing various TVs because the accuracy of classification varies according to the TV. However, determination of the TV is difficult because the optimal TV for the data is not obvious. As a result, a system to automatically determine an appropriate TV for the data is strongly expected. Therefore, a strategy by which the TV is dynamically determined using the same structure of the neural network is developed, as shown in Fig. 2.

Since each function is classified based on whether the EV is greater than the TV, various teacher data are created for each neural network. However, many trials are needed to find the teacher data for the improvement of the classification accuracy. In order to reduce the computation time, an output layer is created for the network, which includes two neural networks, and a network was created to facilitate the creation of teacher data. Figure 3 shows the structure of the output layer of the network.

The following equation can be used to obtain the output value, y, calculated using the sigmoid function:

$$y = \frac{1}{1 + e^{-z}} \tag{1}$$

$$z = 1 * \text{Evaluation value} + (-1) * \text{Threshold value} \tag{2}$$

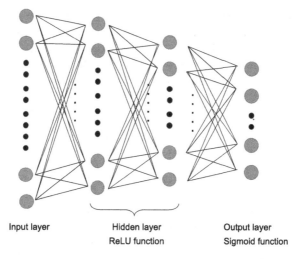

Fig. 2. Structure of the neural network for calculating the EV and the TV

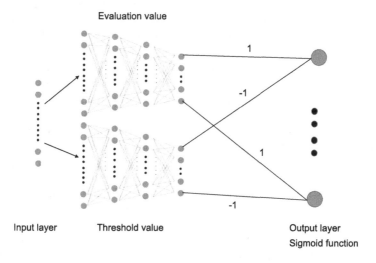

Fig. 3. Structure of the neural network used to classify functions

Here, z in Eq. 2 is a value that indicates whether the EV is greater than the TV. Because each neural network was constructed to output the optimal value by learning only the weight of the network without adjusting coefficients of the EV and the TV, the constant values are used as coefficients. The output value y in Eq. 1 also indicates whether the EV is greater than the TV. The details of the output value are as follows:

$$
\begin{aligned}
0.5 \leq y &\Rightarrow 0 \leq z &\Rightarrow TV \leq EV \\
0.5 > y &\Rightarrow 0 > z &\Rightarrow TV > EV
\end{aligned}
\tag{3}
$$

In Eq. 3, the first line in which y is more than 0.5 has an EV that is greater than or equal to the TV, and the second line in which y is not more than 0.5 has an EV that is less than the TV. Since the second line indicates an unnecessary function for modeling, we adopt the function of the first line. In the present study, a novel method to classify the function called the function classifier is proposed.

4.2   Creation of the Learning and Test Data

Time series data used for learning and testing in the experiment were prepared as follows. Since the experimental data are assumed to be physics data, the time-dependent form is natural for candidate functions. Candidate functions are expressed as $f(t)$, where t is the time and the sampling interval Δt is set to 10^{-3}. We select seven representative functional forms as examples.

$$\log(at + \varepsilon), \ e^{at}, \ \sin(at + b), \ (at + \varepsilon)^{-2}, \ (at + \varepsilon)^{-1}, \ at, \ (at)^2 \qquad (4)$$

Functions $(at + \varepsilon)^{-2}$, $(at + \varepsilon)^{-1}$, at, and $(at)^2$ are examples of algebraic functions. In contrast, functions $\log(at + \varepsilon)$, e^{at}, and $\sin(at + b)$ are examples of transcendental functions. The coefficients a and b are randomly selected and are set to be in the ranges of $-3\sqrt{5} \leq a \leq 3\sqrt{5}$ and $0 \leq b < 2\pi$, where a and b were chosen randomly from uniform random numbers. The value of a ranges between $-3\sqrt{5}$ and $3\sqrt{5}$ because the normal distribution is a simple and effective model to represent complex phenomena. However, if the sampled values are mainly distributed near zero, the complexity of the data will be impaired. Therefore, we changed the method used to generate the distribution of values using uniform random numbers, which can generate approximately the same range of values for a normal distribution with $N(0, 5)$. If the range is three times the standard deviation, statistically, we could reproduce values of the same range. Since b represents the phase of the trigonometric function, the range is set to $[0 , 2\pi]$. There are functions that return infinity when t is zero, but the datum with $t = 0$ is required in order to calculate the derivative. Thus, the small value ε is introduced in order to prevent that the datum from becoming infinity. When calculating derivative values by the numerical differentiation of the second-order precision, the order of the error is $\mathcal{O}(\Delta t^2)$. Therefore, if $\varepsilon = 10^{-6}$, the order of the error is considered not to increase. Data should be created using one or two functions selected randomly from Eq. 4 as follows:

$$\alpha, \beta \sim N(0, 5)$$
$$\alpha \sin\left(2t + \frac{\pi}{2}\right) + \beta \log(0.5t + \varepsilon)$$
$$\alpha(1.5t)^2 \qquad (5)$$

The complexity of the generated data is expressed by coefficients a and b in Eq. 4 and its weight of their linear sum (α, β in Eq. 5). The number of combinations of functions becomes $7 + 21$ patterns, where seven refers to the $_7C_1$ patterns of each function and 21 refers to the $_7C_2$ patterns combining two different functions from seven patterns. For each combination, 100 sets of different coefficients were prepared. Therefore, there are 2,800 sets of data for training and testing.

4.3 Feature Extraction from Data

This section describes the pre-processing for raw data to obtain more accurate results of the estimation. In order to improve the classification accuracy and facilitate learning, the data created in Sect. 4.2 were converted to one-dimensional data from which features were extracted. The one-dimensional data are considered as a pattern like a histogram, and the neural network identifies a function from the pattern.

Feature Extraction by Differentiation. Since differentiation is one method for extracting features from data, the differential value is an appropriate indicator to distinguish a function from other functions. Here, the central difference approximation scheme with second-order accuracy is used for the first, second, third, and fourth derivatives because the second-order central scheme is, in many cases, the basic and proper choice for the finite difference approximation in fluid simulation. The formulas used for the calculation are as follows:

$$\frac{d}{dt}f = \frac{f_{n+1} - f_{n-1}}{2\Delta t} + O(\Delta t^2)$$

$$\frac{d^2}{dt^2}f = \frac{f_{n+1} - 2f_n + f_{n-1}}{\Delta t^2} + O(\Delta t^2)$$

$$\frac{d^3}{dt^3}f = \frac{f_{n+2} - 2f_{n+1} + 2f_{n-1} - f_{n-2}}{2\Delta t^3} + O(\Delta t^2)$$

$$\frac{d^4}{dt^4}f = \frac{f_{n+2} - 4f_{n+1} + 6f_n - 4f_{n-1} + f_{n-2}}{\Delta t^4} + O(\Delta t^2) \tag{6}$$

Since both ends of the data cannot be approximated in the central difference scheme, the differential values at the end points are not used for learning.

Conversion to One-Dimensional Array Data. In order to perform the following eigenvalue analysis to extract features of data, we convert the raw data into one-dimensional array data. We describe the conversion method in detail in Algorithm 1. Since four derivatives, from the first to fourth derivatives, are used, in addition to the function value itself, five data sets can be used. First, standardization is performed on each data set, but at that time, robust statistics, i.e., the median and the normalized absolute central deviation, are used instead of the average or the standard deviation [10]. In addition, the arctan function is applied to the data to make the values the same order of magnitude. Second, two data sets are selected from the five data sets for the data at all time steps, and the eigenvalues of these two data sets are calculated. Since the eigenvalues become a good indicator to express the features of the data, we introduce a minimum of two column vectors, so that the eigenanalysis can be performed. Of course, if we use more column vectors, more features may be extracted. Equation 7 shows the calculation method of the eigenvalues using the longitudinal vector $x_t = (f(t), \dot{f}(t))^{\mathrm{T}}$, where $f(t)$ and $\dot{f}(t)$ indicate the data and the first derivative of the data, respectively.

$$\begin{pmatrix} \lambda_{1,t} & 0 \\ 0 & \lambda_{2,t} \end{pmatrix} = P_t^{\mathrm{T}}(x_t x_t^{\mathrm{T}}) P_t \tag{7}$$

$$P_t = \begin{pmatrix} \cos \theta_t & -\sin \theta_t \\ \sin \theta_t & \cos \theta_t \end{pmatrix}, \tag{8}$$

where λ_t and P_t are the eigenvalue and the rotation matrix at time t, respectively. The rotation matrix, as shown in Eq. 8, is used to diagonalize the matrix $x_t x_t^{\mathrm{T}}$ when the eigenvalue is calculated using Eq. 7. The reason for using the rotation matrix is that θ_t of the rotation matrix is used to calculate the element number of array data, and the eigenvalues are used to calculate the elements of array data. However, $\lambda_{1,t}$ or $\lambda_{2,t}$ becomes zero because $x_t x_t^{\mathrm{T}}$ has only one eigenvalue, since the rank of its matrix is one. The range of θ_t used to compute eigenvalues is $-\pi/4 \leq \theta_t \leq \pi/4$, and the calculation method of the element number is as follows:

$$\text{Element number} = \left\lfloor \frac{\theta + \pi/4}{\pi/2} N \right\rfloor, \tag{9}$$

where N is the number of bins. If θ_t is $\pi/4$, the element number is $N - 1$. The following equation is used to calculate the feature value, which is added to the elements of the array data:

$$\begin{aligned} z_{1,t} &= log(1 + \lambda_{1,t}) \\ z_{2,t} &= log(1 + \lambda_{2,t}) \end{aligned} \tag{10}$$

In Algorithm 1, Nt represents the number of time steps. If the number of bins is four, the one-dimensional array data are calculated as shown in Tables 1 and 2.

Algorithm 1. Calculate One-dimensional Array Data, Array[]

1: Perform standardization
2: $Array[N] = \{0\}$
3: **for** $i = 0$ to Nt **do**
4:    $t = i * \Delta t$
5:    Calculate $\lambda_{1,t}, \lambda_{2,t}, P_t, \theta_t$ by Eqs. 7 and 8
6:    $j1 = \left\lfloor \dfrac{\theta_t + \pi/4}{\pi/2} N \right\rfloor$
7:    $j2 = \left\lfloor \dfrac{-\theta_t + \pi/4}{\pi/2} N \right\rfloor$
8:    $Array[j1] += z_{1,t}$
9:    $Array[j2] += z_{2,t}$
10: **end for**

Table 1. Assigned range of θ for each element

Element number	0	1	2	3
Range of θ	$-\frac{\pi}{4} \leq \theta < -\frac{\pi}{8}$	$-\frac{\pi}{8} \leq \theta < 0$	$0 \leq \theta < \frac{\pi}{8}$	$\frac{\pi}{8} \leq \theta < \frac{\pi}{4}$

In Table 2, the numerical values in the parenthesis indicate the calculated results of the expression in the same cell. After the array data is generated, the data is normalized so that the maximum value is one. An example of the pattern generated using this method is shown in Fig. 4. Since it is possible to calculate 10 ($_5C_2$) types of one-dimensional array data, the one-dimensional array generated by concatenating these data is used as the input layer of the neural network.

4.4   Learning Method

Learning was advanced by the backpropagation method, and the size of the mini batch was set to 100. Mean square error was used as the loss function, and Adam [11] was used to update the layer weights.

In the present study, the original teacher data and the devised teacher data are used. The original teacher data is created so that the output value of the network differs depending on whether each function is used for data creation. The label of a function that is used to create the input data is a "Candidate function", and the label of an unused function is "Not a candidate function". The original teacher data, which is used to classify functions, is shown as Table 3.

If some functions, which are required for modeling, are classified into the "Not a candidate function" category, an appropriate formula will not be obtained by GP. Therefore, the devised teacher data, which can reduce the number of functions, are created in order to increase the possibility of obtaining a formula that expresses the data. Table 4 shows the devised label for the teacher data, and "Output value" indicates the value obtained from the neural network in the learning process.

Table 2. Method of calculating one-dimensional array data

t (time)	θ_t	$z_{1,t}$	$z_{2,t}$	Element 0	Element 1	Element 2	Element 3
0	$\pi/10$	1	0	(0)	$z_{2,0}(0)$	$z_{1,0}\ (1)$	(0)
Δt	$\pi/6$	0	2	$z_{2,\Delta t}(2)$	$z_{2,0}(0)$	$z_{1,0}\ (1)$	$z_{1,\Delta t}(0)$
$2\Delta t$	$-\pi/10$	3	0	$z_{2,\Delta t}(2)$	$z_{2,0}+$ $z_{1,2\Delta t}(3)$	$z_{1,0}+$ $z_{2,2\Delta t}(1)$	$z_{1,\Delta t}(0)$
$3\Delta t$	$-\pi/6$	0	4	$z_{2,\Delta t}+$ $z_{1,3\Delta t}(2)$	$z_{2,0}+$ $z_{1,2\Delta t}(3)$	$z_{1,0}+$ $z_{2,2\Delta t}\ (1)$	$z_{1,\Delta t}+$ $z_{2,3\Delta t}\ (4)$
$\vdots$	$\vdots$	$\vdots$	$\vdots$	$\vdots$	$\vdots$	$\vdots$	$\vdots$

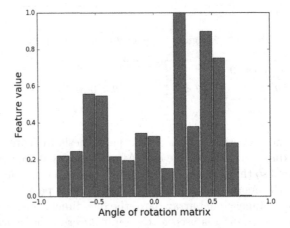

Fig. 4. Histogram pattern of one-dimensional array data representing extracted features. The vertical axis shows the value calculated by Eq. 10

Table 3. Original teacher data

Label	Assigned label
Candidate function	1
Not a candidate function	0

Table 4. Devised teacher data

Label (and conditions)	Assigned label
Candidate function	1
Candidate function && Output value $y < 0.5$	2
Not a candidate function	0

We will describe a method of creating devised teacher data. The combination of teacher data and output value usual belongs to one of the following four types depending on the output value y of the neural network under learning.

1. Assigned label is 1 && the output value $y \geq 0.5$,
2. Assigned label is 1 && the output value $y < 0.5$,
3. Assigned label is 0 && the output value $y \geq 0.5$,
4. Assigned label is 0 && the output value $y < 0.5$.

Although the combination type No. 2 is normally assigned label 1, but the label is changed to label 2 in order to reduce events such as No. 2 . Other combinations use normal teacher data without changing already assigned label. In other words, by changing the teacher data and increasing the error, the goal is to make the network learn so that phenomena such as No. 2 do not occur. In Sect. 5, the difference in the classification accuracy between the original teacher data and the devised teacher data is discussed.

5   Experiment

In this section, the experimental method and the method of evaluating the experimental results, which is the proposed method, are explained for the function classifier. The experimental results are then discussed. The experiment is performed on a PC with the Ubuntu 16.04 LTS operating system and Chainer library version 2.1.0.

5.1   Experimental Method

The length of the one-dimensional array as the input layer of the neural network (NN) is set to 300, and learning and testing are executed. Learning and testing data used in the experiment are created by the method explained in Sect. 4.2. Two experiments were conducted in the present study, and the performances of the function classifier and the devised teacher data were confirmed. The first experiment shows the classification accuracies for different TVs, and the second experiment shows classification accuracies for different teacher data (see Sect. 4.2).

5.2   Evaluation Method

Precision and recall were used for evaluation. Precision is the ratio of functions that can be correctly classified as candidate functions to functions estimated to be candidate functions. Recall is the ratio of functions that can be correctly classified as candidate functions to functions that are actually used for data creation. The confusion matrix, which is used to calculate the precision and the recall, is shown as Table 5.

Table 5. Confusion matrix

		Necessity for modeling	
		Necessary	Unnecessary
Classification result	Candidate	True Positive (TP)	False Positive (FP)
	Not a candidate	False Negative (FN)	True Negative (TN)

Here,

- TP represents the number of functions needed for modeling and that are classified as a "Candidate function",
- FP represents the number of functions that are not needed for modeling and that are classified as a "Candidate function",
- FN represents the number of functions needed for modeling and that are classified as "Not a candidate function",
- TN represents the number of functions that are not needed for modeling and that are classified as "Not a candidate function".

The precision and recall are calculated as follows:

$$precision = \frac{TP}{TP + FP}$$
$$recall = \frac{TP}{TP + FN} \tag{11}$$

5.3   Experimental Results and Discussion

Effect of the Function Classifier. Figure 5(a) and (b) show the changes in the precision and recall according to the proposed method to determine the TV. In the first experiment, learning was conducted using the original teacher data (see Table 3). The green bar represents the classification accuracy using the proposed, and the other bars represent the classification accuracy when functions are classified according to whether the output of the neural network is greater than the given TV.

When the given TV is 0.5, the classification accuracy is the best among the three given TVs. Furthermore, the classification accuracy of the function classifier is as good as that for TV = 0.5. Therefore, the function classifier can obtain the best classification accuracy. The point to emphasize here is that the proposed function classifier achieves high precision and recall without giving the TV value.

Devised Teacher Data. For the second experiment, Fig. 6(a) and (b) show the changes in the precision and recall due to differences in the teacher data. The red and green bars represent the classification accuracies when the original

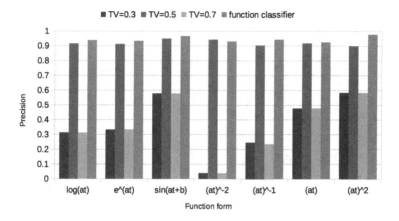

(a) Comparison of precision

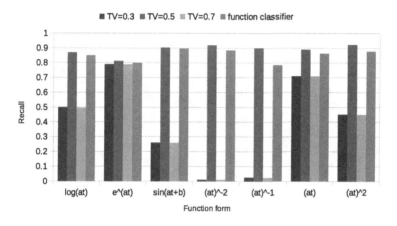

(b) Comparison of recall

Fig. 5. Differences in classification method

teacher data are used. The other bars represent the classification accuracy when the function classifier was learned using the devised teacher data (see Table 4).

From Fig. 6(a) and (b), although the precision decreased when using the devised teacher data, the devised data can improve the recall. In other words, the devised teacher data can effectively reduce the number of FNs, i.e., the number of functions that are needed for modeling but are not used in GP is reduced. When the number of these functions is small, there is a high possibility of preventing the deterioration of the formula obtained using GP. For this reason, if the recall becomes high, it is easier to estimate the correct solution. Therefore, the ingenuity of the teacher data is essential to reduce the computational complexity of GP.

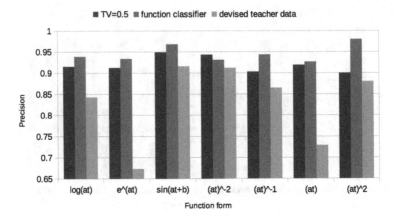

(a) Comparison of precision

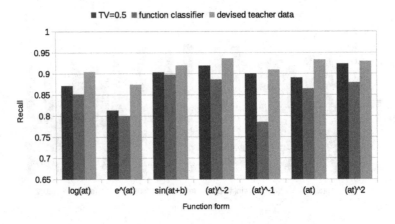

(b) Comparison of recall

Fig. 6. Differences in teacher data

Discussion. In the present study, the prepared data are relatively simple, and noise that may degrade the estimation accuracy is not taken into account. In order to treat practical data, even more complicated data must be classified correctly with high probability, and countermeasures against the noise are necessary. In particular, since the proposed method uses differentiation, it is sensitive to noise. Therefore, there is a risk that the predicted accuracy will be degraded, and methods by which to reduce errors, such as smoothing, should be considered.

6   Conclusion

An effective pre-processing method for solving a symbolic regression problem using the genetic programming was proposed and verified. The proposed method effectively eliminates unnecessary functions from an initial set of functions by a deep neural network that classify whether functions are required. The deep neural network considered in the present study has neural networks for the evaluation value and the threshold value, and the accuracy of classification was improved by introducing an eigenvalue analysis and the results of the two neural networks. The output layer was designed so that the two neural networks can learn simultaneously in order to automatically determine the threshold value, which results in improved classification accuracy. Furthermore, devised teacher data were used to improve the accuracy of classification, and the recall can also be improved. Thus, the proposed method has the ability to effectively reduce the number of candidate functions required for the genetic programming and greatly contributed to reducing the number of computations. The proposed method can successfully classify functions with over 90% of the data created in the present study. Consequently, it was found that it is possible to reduce the computational complexity of the genetic programming using extraction features of data and devised teacher data.

References

1. Schmidt, M.D., Lipson, H.: Distilling free-form natural laws from experimental data. Science **324**(5923), 81–85 (2009)
2. Schmidt, M.D., Lipson, H.: Coevolution of fitness predictors. IEEE Trans. Evol. Comput. **12**(6), 736–749 (2008)
3. Martius, G., Lampert, C.H.: Extrapolation and learning equations. CoRR, arXiv:1610.02995 (2016)
4. Van Hemert, E., Eggermont, J., Hemert, J.I.: Stepwise adaptation of weights for symbolic regression with genetic programming. In: Proceedings of the Twelveth Belgium/Netherlands Conference on Arti Intelligence (BNAIC 2000) (2000)
5. Schmidt, M.D., Lipson, H.: Co-evolution of fitness maximizers and fitness predictors. In: GECCO Late Breaking Paper (2005)
6. Schmidt, M.D., Lipson, H.: Co-evolving fitness predictors for accelerating and reducing evaluations. Genet. Program. Theor. Pract. IV **5**, 113–130 (2006)
7. Uy, N.Q., Hoai, N.X., O'Neill, M., McKay, R.I., Galván-López, E.: Semantically-based crossover in genetic programming: application to real-valued symbolic regression. Genet. Program. Evolvable Mach. **12**(2), 91–119 (2011)
8. Suzuki, I., Ikeuchi, Y., Tsumura, T., Nakashima, Y., Nakashima, H.: A speedup technique for GA with reuse. In: IPSJ Transactions on Advanced Computing Systems, vol. 46, no. SIG 16(ACS 12), pp. 129–143, December 2005
9. Haeri, M.A., Ebadzadeh, M.M., Folino, G.: Statistical genetic programming for symbolic regression. Appl. Soft Comput. **60**, 447–469 (2017)
10. Maronna, R., Lovric, M.: Robust statistical methods. In: Lovric, M. (ed.) International Encyclopedia of Statistical Science, pp. 1244–1248. Springer, Heidelberg (2011). https://doi.org/10.1007/978-3-642-04898-2
11. Kingma, D.P., Ba, J.L.: Adam: A Method for Stochastic Optimization. CoRR, arXiv:1412.6980 (2014)

Toward FPGA-Based Semantic Caching for Accelerating Data Analysis with Spark and HDFS

Marouan Maghzaoui[1], Laurent d'Orazio[2(✉)], and Julien Lallet[1]

[1] Nokia Bell Labs, Nozay, France
{marouan.maghzaoui,julien.lallet}@nokia-bell-labs.com
[2] Univ Rennes, CNRS, IRISA, Lannion, France
laurent.dorazio@irisa.fr

Abstract. With the increase of data, traditional methods of data processing have become time and power inefficient. As enhancement, we propose a new accelerated architecture for querying big Databases. This architecture combines the advantages of the HDFS for the management of huge amount of data and the fast processing of queries of Spark SQL. It also benefits of the processing efficiency of the hardware acceleration of FPGAs and of the semantic caching architecture to process recently used data stored in the cache.

Keywords: FPGA · Spark · HDFS · Semantic caching

1 Introduction

One of the big challenges in the IT sector today is the handling of an ever-growing size of databases. The annual size of data that is created and stored worldwide increases by 31% every year and amount of database computations increases by 21% every year [6].

Processing and analyzing all this data is a big challenge. It requires a mix of processing techniques, data sources and storage formats. Query executions and computations must be accelerated by large rate. Furthermore, large storages must be handled efficiently to manage many read and write operations. This issue is raised in many applications: the Internet, social network, healthcare, cyber security, smart cities, etc. In particular, security monitoring, a domain of cyber security aims to store large amount of logs so as to detect subtle attacks. This require on one hand large storage and on the other hand efficient processing.

Until now, Generic Purpose Processors (GPPs) are used to compute database queries. But GPPs now face the near end of Moore's Law [16], limits in terms of clock frequency [13] and multi-core processing [9]. Hence, we must think of further possibilities to accelerate the query executions.

Recent works [1,2,12] have shown that we could use distributed file systems (DFS) and Massively Parallel Processing (MPP) to store and process huge

© Springer Nature Switzerland AG 2019
D. Kotzinos et al. (Eds.): ISIP 2018, CCIS 1040, pp. 104–115, 2019.
https://doi.org/10.1007/978-3-030-30284-9_7

databases and complex queries. Hadoop Distributed File System (HDFS), the open source version of the google file system GFS [10], and Spark SQL [2] are respectively popular DFS and MPP used by the big data communities. HDFS and Spark rely on GPPs and consequently inherit from GPP limitations mentioned previously. Regarding hardware acceleration, [4,8,14,15,18] have shown that Field Programmable Gate Arrays (FPGAs) could be a better alternative to GPPs in the field of database applications in general and for query-execution purposes with an increasing data throughput with lower energy effort. But not all queries can be easily implemented and processed in a FPGA. It is necessary to have a standard SQL engine to run those non-supported queries. Furthermore, it is reasonable to consider that DFS should be still managed by software. Anyway, intern memory cells of the FPGAs and dedicated RAM memories on board should be used as cache memories to store recent data and results. Semantic caching, which is a technique used for optimizing the evaluation of database queries by caching results of old queries and using them when answering new ones, shows a lot of promise and can be applied to accelerate database queries [17].

This paper describes our effort to create an architecture that combines an accelerated query engine and a semantic cache memory implemented on FPGA, HDFS and Spark to accelerate query executions. This architecture benefits from the hardware acceleration of FPGA, the rapid access to memory in the semantic cache and Big Data capabilities of HDFS and Spark to efficiently execute very large queries.

The remainder of the article is outlined as follows: Sect. 2 discusses background and related work in the field. Section 3 introduces the proposed architecture to accelerate query processing. Finally, a conclusion and some perspectives are given in the last section.

2   Background

This section provides necessary background on Big Data, hardware acceleration and semantic caching. Later, we present recent related research and we discuss their limitations.

2.1   Big Data

According to [5], the term *Big Data* is used to refer to the huge volume of data that are too difficult to store, process and analyze through traditional database technologies like CPU servers in conventional Data Centers within a tolerable time. Some of the solutions used by the Big Data and database communities to handle the problems previously mentioned are Spark SQL and HDFS.

HDFS is a self-healing, distributed file system that provides reliable, scalable and fault tolerant data storage on commodity hardware. HDFS accepts data in any file format like text, images, videos, database regardless of architecture and automatically optimizes high bandwidth streaming. HDFS is designed to hold large amounts of data and provides faster access to data. It is highly scalable anyway limited to 200 Peta Bytes (PB) of storage [3].

Spark is a general-purpose cluster computing engine with libraries for streaming, graph processing, machine learning and SQL. Spark SQL provides a Data Frame API that can perform relational operations on both external data sources and Spark's built-in distributed collections. This API evaluates operations lazily so that it can perform relational optimizations. Also, to support the wide range of data sources and algorithms in Big Data, Spark SQL introduces a novel extensible optimizer [2]. Spark SQL is also a part of Hadoop Ecosystem and can be deployed directly on HDFS so that resources can be statically allocated on some of the machines of an Hadoop cluster.

2.2   Hardware Acceleration

As we reached the end of Moore's law [16], we are experiencing a growing interest in new solutions to boost performance.

The combination of different hardware accelerators, as Graphic Processing Units (GPUs), FPGAs and Application Specific Integrated Circuits (ASICs) along with GPPs, provides a wide panel of solutions from which to choose the most suitable architecture for a specific task. Among these architectures, FPGA provides an excellent acceleration platform. Inherent parallelism, access to local memories as Block Random Access Memories (BRAMs) and the ability to be partially and dynamically reconfigured are interesting characteristics we need to efficiently exploit.

2.3   Semantic Caching

Caching consists in a resource element to accelerate computing processes. Data stored in a cache might be the result of an previous request or a duplicate of recently used data stored elsewhere. Traditional cache architectures are based on page and tuple caching where possible data relationships are not handled efficiently. Semantic caching has been proposed to overcome these drawbacks.

The cache is managed as a collection of semantic regions which are groups of semantically related data. Access and cache replacement are managed at a unit of semantic region. Semantic caching is based on three key features. First, a description of the data stored in the cache is maintained in the form of a compact specification. Requests for missing data in the cache are thus faster fetching to the cache. Secondly, replacement policies are flexible and could be different for each semantic regions, which are associated with collections of tuples. This is to avoid the high overheads of tuple caching and, unlike page caching, is insensitive to bad clustering. Third, maintaining a semantic description of cached data enables the use of sophisticated value functions that incorporate semantic notions of locality, not just Least Recently Used (LRU) or Most Recently Used (MRU) policies in case of cache replacement [7].

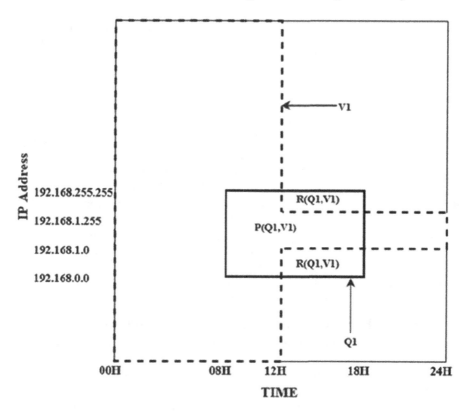

Fig. 1. Example of semantic cache regions

Queries in a semantic cache are split in two parts: a remainder and a probe query. The probe query retrieves the portion of the result already available in the cache. A remainder query fetches any missing data in the cache.

For example, Fig. 1 represents a semantic cache storing log data. The semantic regions are arranged according to the time and IP addresses of the logs. V1 (Eq. 1) represents the result of previous queries (Q_{n-1} (Eq. 2) and Q_{n-2} (Eq. 3) already stored in the cache.

$$V1 = (Time < 12H \lor (IPAddress > 192.168.1.0$$
$$\land IPAddress < 192.168.1.255)) \tag{1}$$
$$Q_{n-1} = (Time < 12H) \tag{2}$$
$$Q_{n-2} = (IPAddress > 192.168.1.0 \land IPAddress < 192.168.1.255) \tag{3}$$
$$Q1 = ((Time < 8H \land Time > 18H) \land$$
$$(IPAddress > 192.168.0.0 \land IPAddress < 192.168.255.255)) \tag{4}$$

A new query Q1 (Eq. 4) would be compared to V1 (V1 $\land$ Q1) to fetch the probe query in the cache (P (Q1, V1) as depicted Fig. 1). The remainder query

(R (Q1, V1)) is then executed to get the missing data. The results of remainder and the probe queries are combined and sent to the user. The result of the remainder query is stored in the cache for future queries.

2.4   Motivation

Static approaches which use FPGAs for query processing [4,8,18] succeeded in implementing query engines capable of doing SQL based processing (select, project join, etc.) with a very high throughput. Some recent research proposes a hybrid architecture (FPGA + CPU) to process LIKE queries [14].

Although some of these existing solutions handled queries for large volumes of data (100 TB in [18]), to the best of our knowledge, the proposed FPGA architectures cannot process huge data stored in multiple servers by the use of a distributed file system. The architecture that we propose, using the Distributed File Systems HDFS, massively parallel processing on Spark SQL, hardware acceleration and semantic caching in FPGA could be the solution to the problem of executing queries in Big Data in a relatively short time.

3   Toward FPGA-Based Semantic Caching for Accelerating Big Data Analysis with Spark and HDFS

In this section, we describe the preliminary building blocks for a FPGA-based semantic caching for Big Data analysis's architecture. First, details of our hardware and software architecture is presented followed by some description of the SQL queries execution. To start with, the system will consider SQL queries with conjunctions only.

In a second part, the FPGA architecture implementation is given. In the following, we detail the management of semantic regions, in particular reads and writes in the FPGA. Analysis and evaluation on the cache entries will be addressed in future works and are out of the scope of this paper.

3.1   Overview

The FPGA-based semantic caching for Big Data analysis's architecture that we propose (Fig. 2) is a hardware software implementation system with a PCI Express interface for inter communication.

The FPGA contains the semantic cache and an accelerated query engine. The software part contains the Query analyzer and Spark. HDFS is the file system where the database is stored.

The Query analyzer contains a catalogue of recent queries and their respective data stored in the semantic cache which is basically the first key idea in semantic caching.

When a user starts a request, the Query analyzer figures out if the requested data is fully stored in the semantic cache (cache hit), partially in the cache or completely missing in the cache (cache miss). The Query analyzer also checks if

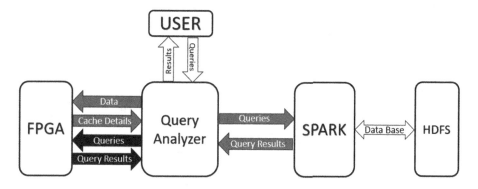

Fig. 2. Architecture of FPGA-based semantic caching for Big Data analysis

the Query engine implemented in the FPGA integrates the query operations to accelerate.

If we have a cache miss or if the query engine cannot process the query (grey arrow in Fig. 2), the query is processed by the Spark query engine. This is the standard, non-accelerated Spark process based on libraries capable of doing complex queries. Spark fetches the data requested in HDFS and executes the query. The data and the results of the query are then sent to the query analyzer. Afterwards, the query analyzer sends the result to the user and proof check if the data and the results can fit in the cache. If they can fit in the cache, the query analyser sends them to the FPGA. When the cache update is done, the FPGA sends cache details to the query analyzer to update the cache catalogue: address of the last data sent and validity of the data cached.

If we have a cache hit and the query engine can process the query (black arrow in Fig. 2), the query is sent to the FPGA. The query engine executes the request and sends the result to the user and the query analyzer. Finally, if the data requested is partially in the cache, the query is parallelized into two or more queries to be executed by Spark and the query engine in FPGA. For example, referring to the example given Sect. 2.3, the query Q1 is divided into a probe query and remainder query. The remainder query is executed by the Spark SQL and the probe query is sent to the FPGA to fetch the data from the semantic cache. Then all the results are sent to the user and remainder query's result are updated in the cache.

3.2 FPGA Structure

The structure in the FPGA (Fig. 3) is composed by three main components: The "Write in Cache", the "Semantic cache" and the "Query engine".

The write in cache is responsible of writing the data to the cache according to the replacement policy chosen. Furthermore, it provides the cache changes to the catalogue stored in the query analyzer. Finally, it is also responsible for managing the semantic regions.

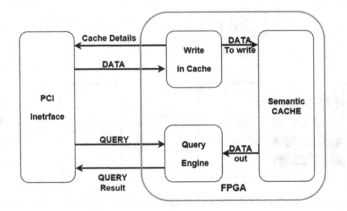

Fig. 3. FPGA structure

The semantic cache memory is a combination of BRAMs linked together to constitute a beta version of a cache memory. In some large FPGAs, the total storage capacity of the BRAMS combined is up to 30 MB. In the future, we plane to extend the cache memory plan to external RAMs to create a second level of functional cache memory.

The query engine reads the query, fetches the necessary data from the cache, executes the query and sends the results to the query analyzer. The query engine must be capable of executing as much different queries as possible, reading and storing different formats of databases which is a real challenge for FPGA acceleration.

4   Experiments

In this section, we introduce the implementation details of the FPGA prototype. Then, we provide some performance results of the prototype and compare them to non accelerated standard Spark SQL.

4.1   Implementation

The prototype is composed by a server for software processing and an FPGA board for hardware processing. The intercommunication between both is managed by RIFFA [11]. RIFFA is a simple framework for communicating data from a host Central Processing Unit (CPU) to a FPGA via a PCIe. RIFFA communicates data using direct memory access (DMA) transfers and interrupt signalling. This achieves high bandwidth over the PCIe. Our implementation relies on an XpressV7-LP HE design board based on a Xilinx Virtex-7 FPGA XC7VX690T. The server used for the evaluation is based on an Intel R Xeon R CPU E5-1620 able to run at 3.60 GHz with 16 GB of RAM. The operating system is a Linux Ubuntu 16.04.4 Long Term Support distribution based on kernel 4.4.0-119-low-latency.

Table 1. FPGA resource utilization

Resource	Utilisation	Available	Utilisation %
LUT	21825	433200	5.04
LUTRAM	322	174200	0.18
FF	26741	866400	3.09
BRAM	186.5	1470	12.69
IO	7	600	1.17
GT	8	20	40
BUFG	4	32	12.5
MMCM	1	20	5
PCIe	1	3	33.33

Table 1 shows that the prototype is well optimized with limited resources footprint (5% of Look Up Tables (LUTs), 3% of Flip Flops (FFs) and 12% of Block Random Access Memories (BRAMs)).

4.2   Results

To test the prototype, we tested the upload speed and the download speed. Then we tested different queries and compared them to standard non-accelerated Spark.

To test the upload speed, we sent four times different sizes of data to the FPGA (from 16 Bytes to 384 KB) and got the average to have a more reliable result. Figure 4 presents the results on upload throughputs. The speed of upload which can be achieved is above 1 GB/s when the size of the data is more than 256 KB.

To test the download speed, we sent four times read queries to read different sizes of data from the FPGA (from 16 Bytes to 384 KB) and got the average to have a more reliable result. Figure 5 presents the results on download throughputs. The speed of download which can be achieved is above 1 GB/s when the size of the data is more than 256 KB.

To test the query speed, we sent four times different queries for different sizes of data stored in the FPGA caches (from 16 Bytes to 384 KB) and got the average time to have a more reliable result. The different queries are:

- Simple "where" query "SELECT * FROM tableName WHERE condition" where the condition is valid for all the data (returns all).
- Simple "where" query "SELECT * FROM tableName WHERE condition" where the condition is not valid for all the data (returns ∅).
- Simple "where" query "SELECT * FROM tableName WHERE condition1 and condition 2 and condition 3 and condition 4" where the conditions are valid for all the data (returns all).

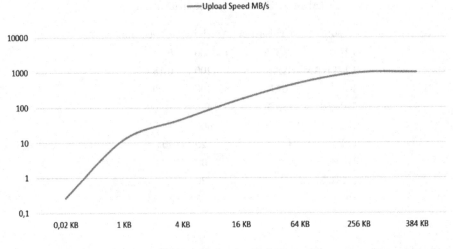

Fig. 4. Upload speed test

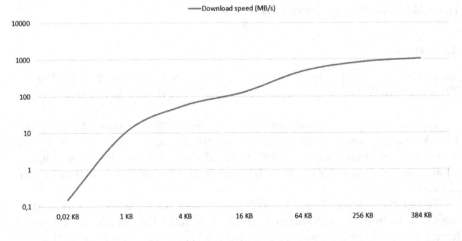

Fig. 5. Download speed test

– Simple "where" query "SELECT * FROM tableName WHERE condition1 and condition 2 and condition 3 and condition 4" where the conditions are not valid for all the data (returns ∅).

Figure 6 shows that the fastest query to execute is the simplest query with no valid condition while the slowest is the complex query which returns all data. This can be explained by two reasons. The first is that Query Analyzer takes more time to analyze a complex Query (between 170 μs and 140 μs) than a simple query (between 100 μs and 120 μs). The second reason is that the FPGA takes time to send Data back to the software layer when Data are present in the

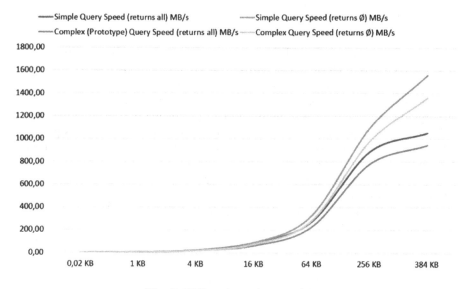

Fig. 6. Different queries speed test

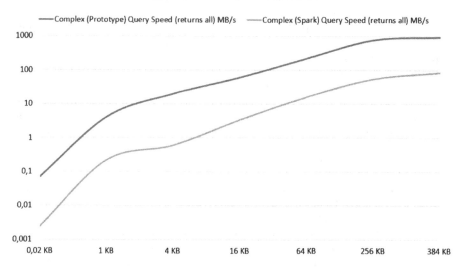

Fig. 7. Complex queries speed comparison

FPGA cache. The same figure also provides a range of throughput results for Data sizes above 256 KB.

Figure 7 presents the results of throughput comparison between Spark and the prototype. First, we did a comparison between the slowest queries and complex queries that returns all data requested. We can notice that the prototype is about 10 times faster on average than Spark.

5   Conclusion

In this paper, we proposed an FPGA-based architecture for the acceleration of Big Databases SQL querying. The hardware section is composed of FPGA based query engine and semantic caching. While the software section is composed of a Query analyzer, Spark SQL and the Hadoop Distributed File System (HDFS). This architecture could be later improved by introducing partial reconfiguration to the FPGA architecture to have the ability to do as much queries as possible by the query engine like what was done in [8,18]. This architecture could also be improved by adding more than one FPGA in the architecture to have different FPGAs executing different queries at the same time. In addition, future works include extending the proposed architecture to increase parallelism and distribution. In particular two cases will be considered: (1) edge computing and (2) cloud federations.

References

1. Soomro, T.R., Shoro, A.G.: Big data analysis: Apache spark perspective. Glob. J. Comput. Sci. Technol. (2015)
2. Armbrust, M., et al.: Spark SQL: relational data processing in spark. In: SIGMOD International Conference on Management of Data, Melbourne, Victoria, Australia, pp. 1383–1394 (2015)
3. Bansod, A.: Efficient big data analysis with apache spark in HDFS. Int. J. Eng. Adv. Technol. (IJEAT) 4(6), 313–316 (2015)
4. Becher, A., Ziener, D., Meyer-Wegener, K., Teich, J.: A co-design approach for accelerated SQL query processing via FPGA-based data filtering. In: International Conference on Field Programmable Technology (FPT), Queenstown, New Zealand, pp. 192–195 (2015)
5. Chen, M., Mao, S., Liu, Y.: Big data: a survey. Mobile Netw. Appl. (MONET) 19(2), 171–209 (2014)
6. Cisco Global Cloud: Cisco global cloud index: Forecast and methodology, 2016–2021 white paper. Technical report, Cisco (2010)
7. Dar, S., Franklin, M.J., Jónsson, B.T., Srivastava, D., Tan, M.: Semantic data caching and replacement. In: International Conference on Very Large Data Bases (VLDB), Mumbai (Bombay), India, pp. 330–341 (1996)
8. Dennl, C., Ziener, D., Teich, J.: On-the-fly composition of FPGA-based SQL query accelerators using a partially reconfigurable module library. In: International Symposium on Field-Programmable Custom Computing Machines (FCCM), Toronto, Ontario, Canada, pp. 45–52 (2012)
9. Esmaeilzadeh, H., Blem, E.R., Amant, R.S., Sankaralingam, K., Burger, D.: Dark silicon and the end of multicore scaling. IEEE Micro 32(3), 122–134 (2012)
10. Ghemawat, S., Gobioff, H., Leung, S.-T.: The Google file system. In: Symposium on Operating Systems Principles (SOSP), Bolton Landing, NY, USA, pp. 29–43 (2003)
11. Jacobsen, M., Richmond, D., Hogains, M., Kastner, R.: RIFFA 2.1: a reusable integration framework for FPGA accelerators. ACM Trans. Reconfig. Technol. Syst. 8(4), 22:1–22:23 (2015)

12. Manikandan, S.G., Ravi, S.: Big data analysis using Apache Hadoop. In: International Conference on IT Convergence and Security (ICITCS), Beijing, China (2014)
13. Ross, P.E.: Why CPU frequency stalled. IEEE Spectr. **45**(4), 72 (2008)
14. Sidler, D., István, Z., Owaida, M., Kara, K., Alonso, G.: doppioDB: a hardware accelerated database. In: International Conference on Management of Data, SIGMOD Conference 2017, Chicago, IL, USA, pp. 1659–1662 (2017)
15. Teubner, J.: FPGAs for data processing: current state. Inf. Technol. (IT) **59**(3), 125 (2017)
16. Theis, T.N., Wong, H.P.: The end of Moore's law: a new beginning for information technology. Comput. Sci. Eng. **19**(2), 41–50 (2017)
17. Vancea, A., Stiller, B.: CoopSC: a cooperative database caching architecture. In: 2010 International Workshops on Enabling Technologies: Infrastructures for Collaborative Enterprises (WETICE), Larissa, Greece, pp. 223–228 (2010)
18. Ziener, D., et al.: FPGA-based dynamically reconfigurable SQL query processing. ACM Trans. Reconfig. Technol. Syst. (TRETS) **9**(4), 25:1–25:24 (2016)

A First Experimental Study
on Functional Dependencies
for Imbalanced Datasets Classification

Marie Le Guilly[(⊠)], Jean-Marc Petit, and Marian Scuturici

Université de Lyon, CNRS, INSA-LYON, LIRIS, UMR5205,
69621 Villeurbanne, France
{marie.le-guilly,jean-marc.petit,marian.scuturici}@insa-lyon.fr

Abstract. Imbalanced datasets for classification is a recurring problem in machine learning, as most real-life datasets present classes that are not evenly distributed. This causes many problems for classification algorithms trained on such datasets, as they are often biases towards the majority class. Moreover, the minority class often yields more interest for data scientist, when at the same time it is also the hardest to predict. Many different approaches have been proposed to tackle the problem of imbalanced datasets: they often rely on the sampling of the majority class, or the creation of synthetic examples for the minority one. In this paper, we take a completely different perspective on this problem: we propose to use the notion of *distance* between databases, to sample from the majority class, so that the minority and majority class are as distant as possible. The chosen distance is based on functional dependencies, with the intuition of capturing inherent constraints of the database. We propose algorithms to generate distant synthetic datasets, as well as experimentations to verify our conjecture on the classification on distant instances. Despite the mitigated results obtained so far, we believe this is a promising research direction, at the intersection of machine learning and databases, and it deserves more investigations.

1 Introduction

Databases, machine learning and data mining, are very important domains in computer science. With the recent context of *Big Data*, they are receiving an increased interest and experiencing an important expansion, both in academia and industry. However, as important as they all are, they have tended to grow in parallel, with limited interactions with one another.

There is everything to gain in bringing those domains together, and combining them to propose innovative solutions: this argumentation was already defended in 1996 in [14], and has since been followed in recent works trying to increase the permeability between these domains. Among those, some have tried to make use of machine learning algorithms to improve the use of databases, with query discovery [24], query inference [6], or query rewriting [9]. Others are trying to integrate machine learning into databases such as [26] or [10].

© Springer Nature Switzerland AG 2019
D. Kotzinos et al. (Eds.): ISIP 2018, CCIS 1040, pp. 116–133, 2019.
https://doi.org/10.1007/978-3-030-30284-9_8

On the one hand, data dependencies are a powerful notion, that has been thoroughly studied for both relational databases and data mining. It has been applied in various situations, from the discovery of data quality rules [8] to association rule mining [2]. On the other hand, supervised classification is a long studied problem in machine learning.

This paper investigates whether or not those two independent domains can benefit from one another. Indeed, existing supervised learning solutions [19] do not seem to make any use of data dependencies. Is there a reason for this, and is it possible to find supervised classification problem that would be suited for the use of data dependency?

Data dependencies, and more specifically functional dependencies, give global constraints on a dataset: they describe constraints between attributes, and in some way describe the structure of a given relation. If two relations are defined on the same schema, they might satisfy similar dependencies, or even the exact same one, or on the opposite completely different ones: quantifying this difference of functional dependencies, with some sort of metric or distance, is then a way to evaluate the similarity of two relations in terms of structure and underlying patterns. With such a distance, which is then more semantic than "traditional" distances (e.g. euclidean, Manhattan, ...), interesting sub-problems of supervised classification could be addressed with a new approach. In particular, it can be interesting to look at the problem of imbalanced datasets: this happens in binary classification, when one class is much bigger than the other. In this situation, classifiers are biased and tend to always predict the majority class as there are many more examples of it. But often the minority class is actually much more interesting and is the one that data analysts want to predict accurately. One example of this is a dataset of banking transactions, which only contains a tiny proportion of fraudulent transactions, against thousands of regular ones. But this tiny portion is still much more interesting as they are the one that are crucial to detect!

Dealing with imbalanced datasets is a common issue in machine learning, and various strategies have been proposed [18]. Among those, a common one is the undersampling of the majority class: select only a subset of examples in it, such that its size is similar to the one of the minority class. There are several undersampling strategies: basic random sampling, identifying clusters in the majority class, ... But to the best of our knowledge, there is no technique that makes use of global constraints for imbalanced datasets. Using the notion of a distance based on data dependencies, and selecting *distant* tuples (or on the contrary *close* ones), makes sense: such an undersampling strategy could provide a disruptive approach to the imbalanced dataset problem, by not only looking at data values but also at the structural properties of data.

The general idea is to compute a set of functional dependencies satisfied by the minority class, and compute the set of functional dependencies that are as distant (or close) as possible from it.

The challenge is then to identify the tuples in the majority class that, together, satisfy exactly (or as many as possible) this set of dependencies. This is not a trivial problem, and it raises several combinatorial challenges. Moreover, it is based on this notion of semantic distance based on data dependencies: it makes the study of such a problem really ambitious but also quite perilous. As a first step, it is therefore possible to consider several subproblems, that can be summarized as follow:

Is it easier to classify imbalanced datasets between *distant* relations? And can functional dependencies help to identify better balanced classification datasets?

In this paper, we present our ongoing work, to partially answer these two difficult questions. Our approach works as follows: instead of getting dependencies from existing data, we propose to first determine the required constraints in order to then generate synthetic data accordingly. We can then refine our problem statement:

Is it possible to find synthetic datasets verifying this conjecture: datasets that are distant in terms of DF are easier to classify?

To answer this, the contributions of this paper are then:

- The use of a semantic distance based on functional dependencies and closure systems, as described in [17].
- The construction of a synthetic imbalanced dataset such that the distance between the minority and the majority class is maximum.
- Experimentations, applying various classification models, to compare classifiers performances when discriminating between the different relations generated.

The purpose is first to point out if synthetic relations generated as *distant* are easier to classify than random relations. Then, further experimentations verify how well classifier trained on such relation adapt when tested on the imbalanced dataset. The obtained results are mitigated, reflecting the difficulty of the new problem presented in this paper. Even if further investigations will be necessary in future works, especially regarding the values of the synthetic data, we believe that we propose a first attempt at tackling an important problem at the intersection of machine learning and databases.

Section 2 introduces the necessary preliminary notions required to understand the paper. Then Sect. 3 explicits the notion of distance between instances of a database, and explains the intuitions lying behind it. Section 4 describes the generation strategies, from closure systems to imbalanced dataset. Based on this, experimentations are detailed in Sect. 5, comparing different approaches, and testing the conjecture of this paper on various classification algorithms. Finally, Sect. 6 offers a discussion on the presented approach and related works, as well as a conclusion.

2   Preliminaries

We first recall basic notations and definitions that will be used throughout the paper. It is assumed that the reader is familiar with databases notations (see [20]).

Let U be a set of attributes. A relation schema R is a name associated with attributes of U, i.e. $R \subseteq U$. A database schema $\mathcal{R}$ is a set of relation schemas.

Let D be a set of constants, $A \in U$ and R a relation schema. The domain of A is denoted by $dom(A) \subseteq D$. A tuple t over R is a function from R to D. A relation r over R is a set of tuples over R. The active domain of A in r, denoted by $ADOM(A, r)$, is the set of values taken by A in r. The active domain of r, denoted by $ADOM(r)$, is the set of values in r.

We now define the syntax and the semantics of a Functional Dependency (FD).

Let R be a relation schema, and $X, Y \subseteq R$. A FD on R is an expression of the form $R : X \to Y$ (or simply $X \to Y$ when R is clear from context).

Let r be a relation over R and $X \to Y$ a DF on R. $X \to Y$ is satisfied in r, denoted by $r \models X \to Y$, if and only if for all $t_1, t_2 \in r$, if $t_1[X] = t_2[X]$ then $t_1[Y] = t_2[Y]$.

Many concepts have been defined on set of FDs, some of them are presented below.

Let F be a set of FDs on U and $X \subseteq U$. The closure of X w.r.t F, denoted by X_F^+, is defined as: $X_F^+ = \{A \in U \mid F \models X \to A\}$ where $\models$ means "logical implication". X is closed w.r.t F if $X_F^+ = X$. The closure system $CL(F)$ of F is the set of closed sets of F: $CL(F) = \{X \subseteq U | X = X_F^+\}$

There exists a unique minimal subfamily of $CL(F)$ irreducible by intersection, denoted by $IRR(F)$, and defined as follows:

- $IRR(F) \subseteq CL(F)$, $U \notin IRR(F)$
- for all $X, Y, Z \in IRR(F)$, if $X \cap Y = Z$, then $X = Z$ or $Y = Z$.

Finally, the concept of Armstrong relations [3] allows to obtain relations satisfying a set of functional dependencies, and only those dependencies. Let F be a set of FD on R. A relation r on R is an Armstrong relation for F if $r \models X \to Y$ if and only if $F \models X \to Y$.

There exists a relationship between Armstrong relations and closure systems [4]. First, agree sets have to be defined. Let r be a relation over R and $t_1, t_2 \in r$. Given two tuples, their agree set is defined as: $ag(t_1, t_2) = \{A \in R | t_1[A] = t_2[A]\}$. Given a relation, their agree sets are then: $ag(r) = \{ag(t_1, t_2) | t_1, t_2 \in r, t_1 \neq t_2\}$.

Then, the relationship can be given as a theorem [4]:

Theorem 1. *Let F be a set of FDs on R and r be a relation over R. r is an Armstrong relation for F if and only if $IRR(F) \subseteq ag(r) \subseteq CL(F)$.*

3   Distance Between Databases

The notion of distance has been studied for years. In computer science, distances are often required, especially in machine learning, for example for Clustering algorithms or K-nearest-neighbors (see [13]). It is possible to define distances between numerical values, vectors, but also words, sentences, ... However, the notion of distance between databases is not a notion that seems to have been given much attention. A first attempt can be found in [21], that proposes an update distance between databases, similarly to the edit distance for strings: the distance between two databases is the minimal number of modification operations to be applied to one database to obtain the other one. However, this distance is not symmetric, and is mostly defined for cases of multiple replications of a database, when the same database is duplicated and modified at different places.

Another definition of distance between databases is given in [17]. This distance is defined in terms of functional dependencies, using the notion of closure systems. Formally, the distance between two databases, instances of the same schema, is defined as follows:

Definition 1 [17]. Let r_1 and r_2 be two relations over $\mathcal{R}$, and F_1 (respectively F_2) the FDs statisfied in r_1 (respectively r_2). The distance between r_1 and r_2 is:

$$d(r_1, r_2) = |CL(F_1) \triangle CL(F_2)|$$

where $A \triangle B$ denotes the symmetric difference of the two sets, i.e:

$$A \triangle B = A \setminus B \cup B \setminus A.$$

This definition can be puzzling at first, and it is necessary to understand the intuitions that lie behind it. Closure systems are intimately related to functional dependencies: therefore similar closure systems lead to similar FDs, and vice-versa. As a consequence, regardless of the values taken by the values of the instances, this distance characterizes the similarity of two sets of functional dependencies. Indeed, distant relations tend to have opposite FDs, while close ones will have similar or even shared FDs.

Moreover, the following property holds:

Property 1 [17]. Let $|\mathcal{R}| = n$. Then $d(r_1, r_2) \leq 2^n - 2$ for any two instances of schema $\mathcal{R}$.

Example 1. Let's take the two following closure systems:

- $CL_1 = \{ABC, AB, AC, A, \emptyset\}$
- $CL_2 = \{ABC, BC, B, C, \emptyset\}$

For two relation r_1 and r_2 with respective closure systems CL_1 and CL_2, the distance is:

$$d(r_1, r_2) = |\{AB, AC, A, BC, B, C\}| = 2^3 - 2 = 6$$

r_1 is a far as possible from r_2 and vice-versa. Moreover, they respect the following sets of functional dependencies:

- $r_1 \models \{B \rightarrow A, C \rightarrow A\}$
- $r_2 \models \{A \rightarrow BC\}$

This illustrates this intuition of "opposites" functional dependencies in distant relations.

This notion of distance between databases is clearly semantic: it does not look at all at the domains of the relation's attributes, or at their types. Instead, it captures underlying patterns and structure, through functional dependencies, and evaluates the distance between relations as the similarity in terms of such structure and patterns between them.

This is exactly the semantic distance that gives an indication of how two relations are distant regarding their functional dependencies. The strong semantic meaning carried by this distance could match the needs required for the problem defined in this paper as motivated in the introduction.

4   Imbalanced Datasets Generation

In order to verify the conjecture that distant relations, in terms of functional dependencies, should be easier to classify, a five step process is proposed. The idea is to recreate the conditions of an imbalanced dataset, simulating two different sub-samples r^+ and s from a majority class Z, to compare their classification against a minority class r^-. More specifically, we simulate an imbalanced dataset, where r^+ and s are two different samples of Z, generated with two different strategies: tuples in s are selected at random from Z, while r^+ is generated to be as distant from r^- as possible. The structure of those different relations is outlined on Fig. 1.

We follow the following process:

Step 1: For a given schema $\mathcal{R} = \{A_1, ..., A_n\}$ of size n, create two closure systems CF^+ and CF^-, as close as possible in size, but as different as possible in terms of attribute set.

Step 2: From step 1, generate two Armstrong relations r^+ and r^- for CF^+ and CF^-.

Step 3: Apply a classification model to discriminate between r^+ and r^-.

Step 4: Create a synthetic majority relation Z over $\mathcal{R}$, from which r^+ is supposed to be a sample.

Step 5: Sample Z to get s, such that $|s| = |r^-|$, and classify between s and r^-.

The objective is to see if the performances of classifiers are better when discriminating between r^+ and r^-, two relations constructed using "opposite" functional dependencies, than between $|s|$ and $|r^-|$, where s is "only" a random sample. This approach raises several questions and sub-problems, that are addressed in the following sections. As expected, the most crucial one is the data values taken to build r^-, r^+ and therefore Z. Indeed, by definition, FDs care about data equality, but independently of the data values themselves. However, the classification algorithms do care about them, and therefore the generation strategy might modify the results obtained in our experiments. It is therefore important to approach this problem carefully.

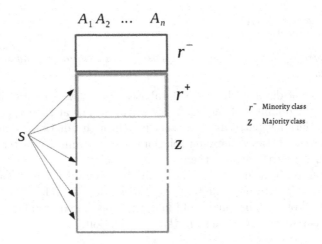

Fig. 1. Generation process for random data

4.1   Synthetic Closure Systems Generation

The first problem is to generate two closure systems such that (1) the size of their symmetric difference is maximum and (2) the sizes of the two closure systems is as close as possible. This is not a trivial problem: given a schema $\mathcal{R}$ of size n, many different closures system can be obtained satisfying this condition.

This problem is interesting and difficult, but is not the center of our paper. To be able to generate automatically two closure systems with a maximized symmetric difference, we propose Algorithm 1. It uses a level-wise (top-down) breadth-first approach strategy on $\mathcal{P}(\mathcal{R})$:

The algorithm works as follows:

– The two closure systems are initialized with $\mathcal{R}$ as it necessarily belongs to each of them.
– At a given level, all elements that do not belong in either CF^- or CF^+ are considered as available candidates for insertion in one of the closure systems. They are selected in a random order, so that each execution of the algorithm does not ensure to produce the same result. This way, we can obtained various closure systems to work with.
– Before inserting an element e, it is necessary to verify if its insertion would satisfy the properties of a closed set. Thus, if the intersection of an element e with any of the elements of same size in CF^- is an element from CF^+, then e has to be added to CF^+, and vice-versa.
– Otherwise an element is added to the closed set with the smallest number of elements. This is to obtain the closest possible size between CF^- and CF^+.
– Once it is decided in which set an element is added, it is also necessary to add all its intersections with elements of same size that already are in the closure system: this is the role of the recursive function **add**.

Algorithm 1. Closure systems generation procedure

---

1 procedure ClosureSystems $(\mathcal{R})$;

    Input : A schema $\mathcal{R}$

    Output: Two closure systems CF^- and CF^+ such that
$$CF^- \triangle CF^+ = 2^{|R|} - 2$$

2 $CF^- = \{R, \emptyset\}$

3 $CF^+ = \{R, \emptyset\}$

4 **for** $l = |\mathcal{R}|$ *to* $l = 1$ **do**

5     $available = \{e \in \mathcal{P}(\mathcal{R})| \ |e| = l, \ e \notin CF^- \text{ and } e \notin CF^+\}$

6   **for** *each e in available* **do**

7   **if** $\exists a \in inter(e, CF^-)$ *such that* $a \in CF^+$ **then**

8          $add(CF^+, e)$

9   **end**

10   **else if** $\exists a \in inter(e, CF^+)$ *such that* $a \in CF^-$ **then**

11          $add(CF^-, e)$

12   **end**

13   **if** $e \notin CF^-$ *and* $e \notin CF^+$ **then**

14   **if** $|CF^-| < |CF^+|$ **then**

15           $add(CF^-, e)$

16   **end**

17   **else**

18           $add(CF^+, e)$

19   **end**

20   **end**

21   **end**

22 **end**

23 **return** $CF1, CF2$

---

24 **Function** add(CF, e):

25     $sameSize = \{a \in CF | length(a) = length(e)\}$ **for** $a \in sameSize$ **do**

26       $i = a \cap e$

27   **if** $i \notin CF$ **then**

28         $add(CF, i)$

29   **end**

30   **end**

The idea behind this algorithm is, given a schema R, to divide evenly all elements from $\mathcal{P}(\mathcal{R})$ into the closure systems. Therefore, they can not be constituted at random, and the insertion of an elements in a closed set has to guaranty some properties, especially regarding the closure by intersection. This algorithm allows to obtain diverse closure systems even for schema of consequent size, automatizing a task which is not feasible "manually". This turns out to be valuable for experimentations, as various closure systems, and therefore various relations, can be tested using this algorithm. It should however be noted that this algorithm has an exponential complexity in the size of $\mathcal{R}$, that limits the size of schema that can be used, if the closure systems are to be obtained in a reasonable amount of time. In practice, we set $|\mathcal{R}| = 12$ in our experimentations.

4.2   Data Generation from Closure Systems

Once two closure systems are generated it is quite easy to derive relations from them, using Armstrong relations, as explained in [3]. Indeed, the structure of an Armstrong relation for a set of functional dependencies is a problem that as already been addressed (see [4]). It relies on the results of Theorem 1.

Given a closure system, an Armstrong relation is defined with a reference tuple t_0 to encode each member of the irreducible set obtained from the closure system with respect to this reference: for a tuple t encoding the ith element X, $t[X] = t_0[X]$ and $t[A] = i$, with $A \in \mathcal{R} \setminus X$.

Example 2. *Let's take the first closure systems from Example 1:*

$CF^- = \{ABC, AB, AC, A, \emptyset\}$ *and* $IRR^- = \{ABC, AB, AC\}$

Relation r^- is derived from CF^-:

r^-	A	B	C	encodes
	0	0	0	reference
	0	0	1	AB
	0	2	0	AC

In our setting, the situation is slightly more complicated, as there are two closure systems. Therefore it implies to make some additional choices regarding the generation of relations, especially the second one. The future use of those relations, i.e for classification problems, should also be taken into account.

Indeed, there is a balance to be found, in order to obtain a convincing classification dataset: the dataset should have two classes, so that their values are not completely similar, but also such that the problem to solve is not trivial, meaning the two classes should overlap in some way.

In particular, there is the question of the reference value for the second relation: it could be the same for both of them. But that would mean that the two relations would have one tuple in common, and a great number of similar tuples, as they would use the same values and therefore have the exact same active domain. Moreover, if in one relation, all the tuples share the same reference value, the classification problem might become too trivial, as the algorithm might be biased toward learning weather or not this specific reference value appears in a tuple, and therefore discriminate between the classes solely based on their respective reference value.

This balance is a crucial point of our process, that deserves more investigations in future works. In this paper, we propose two elements to address this problem:

- For one relation, the reference value to use is based on the previous tuple that was inserted in the relation.
- The two relations do not share any common reference value

We formalize our approach in Algorithm 2, that details how the relations are created in order to respect the given constraints.

Algorithm 2. Relations generation given two closure systems

---

1 procedure RelationsFromCS ($\mathcal{R}, CF^-, CF^+$);

Input   : A schema $\mathcal{R}$, two closures systems CF^- and CF^+

Output: Two relations r^- and r^+ with respective closure systems CF^- and CF^+ and overlapping active domains

2 $n = |\mathcal{R}|$

3 $values = [0...2^n + 2]$

4 $r^- = $ ArmstrongRelation(CF^-, values, n)

5 $r^+ = $ ArmstrongRelation(CF^+, values, n)

6 **return** r^-, r^+

7 **Function** ArmstrongRelation($CF, values, n$):

8     $r \leftarrow |\mathcal{R}| * |CF|$ matrix

9   refvalue $= $ random(values)

10   values.remove(refvalue)

11    $t_{ref} = [\text{refvalue}] * n$

12    $r[0] = t_{ref}$

13   i = 1

14   **for** *each* $e \in CF$ **do**

15   randomvalue $= $ random(values)

16   values.remove(randomvalue)

17       $t = [\text{randomvalue}] * n$

18   **for** *each* $X \in e$ **do**

19         $t[X] = r[i-1][X]$

20   **end**

21       $r[i] = t$

22       $i + +$

23   **end**

24   **return** r

It works as follows:

- Given a schema $\mathcal{R}$ of size n, a pool of possible values is generated, with all integers values from 0 to $2^n + 2$, which is equal to $|CF^-| + |CF^+|$ plus two reference values. This represents all the values that have to be used to generate the two Armstrong relations.
- Then function **ArmstrongRelation** is applied to CF^- to generate r^-.
- A reference value is selected at random in the pool of possible values. It is used to construct the reference tuple, so it is then no longer a possible new value (line 10).
- Then for each element in the closure system, a random value is selected (at random) in the pool of remaining possible values. It is used to create a new

tuple, and the random value is thus removed from the pool. Each attribute from the considered element is encoded with respect to the previous tuple in the relation.

- Once r^- is complete, the same procedure applies for r^+, using the remaining values in the pool of available values.

Example 3. *Following Example 2, and using the second closure system from Example 1:*

$$CF^+ = \{ABC, BC, B, C, \emptyset\} \text{ and } IRR^+ = \{ABC, BC, B, C\}$$

Applying Algorithm 2 we could obtain:

r^+	A	B	C	encodes
	3	3	3	*reference*
	7	3	3	BC
	2	3	2	B
	9	9	2	C

r^-	A	B	C	encodes
	1	1	1	*reference*
	1	1	6	AB
	1	4	6	AC
	1	5	5	A

4.3   Random Data Generation

Once the two relations derived from closure systems are generated, it is time to generate relations Z and s, in order to give a basis for comparison of classifiers. The generation of such a relation Z has to be thought carefully, and be coherent with the generation technique previously applied. For sake of clarity, we only consider values in $\mathbb{N}$.

The final objective is to study the case of imbalanced dataset, and to propose a new undersampling technique based on functional dependencies. The active domain of this relation is crucial, and various strategies are possible:

- $ADOM(Z) = ADOM(r^+)$
- $ADOM(Z) = ADOM(r^+ \cup r^-)$
- $ADOM(Z) = ADOM(r^+ \cup r^-) \cup I$ where I is an interval in $\mathbb{N}$.

The last solution has some flaws, because it would add tuples with values in I that would be easier to classify, with the same problem than explained in Example 3: better classification score would then not be due to functional dependencies but to the fact that tuples are on two distinct domains that are easy to separate. But the two first possibilities are both interesting, giving two possible levels of difficulty. The first one, $ADOM(Z) = ADOM(r^+)$, proposes the exact same conditions when comparing between r^- versus r^+ and r^- versus s: indeed both r^+ and s have an active domain that is overlapping with r^-'s one, but with no common value. The second possibility, $ADOM(Z) = ADOM(r^+ \cup r^-)$, introduces some noise as s and r^- could have values in common, which would provide a more challenging dataset. Therefore, these two options are explored in the experimentations (Sect. 5).

Once the active domain of Z is set, its generation is not complicated, and the number of possible tuples is bounded.

Property 2. *Let $\mathcal{R}$ be a schema of size n, and Z a relation such that $|ADOM(Z)| = m$. The number of possible tuples for Z is the number of permutations of size n from an alphabet of m elements, i.e $max(|Z|) = m^n$.*

This grows considerably fast, and clearly the size of Z can be arbitrarily large. Generating all possibilities would take time and consume memory. Therefore, to limit the size of Z, we consider the factor sf, which is the ratio of size difference between r^- and Z:

$$sf = \frac{|Z|}{|r^-|}$$

This allows to adapt the "imbalanceness" of the dataset, and to generate only a limited number of tuples for Z: the generation method for this relation is simply to create a relation on a schema R of size n, with a fixed number of $|r^-| * sf$ tuples, and for each tuple and each attribute, select randomly a value in $ADOM(Z)$.

Finally, relation s is just a random sample of size $|r^-|$ over $Z \cup r^+$, just like a random undersampling of the majority class for an imbalanced dataset problem.

4.4   Classification Problem

At this point, all necessary relations have been generated. We have the minority class r^-, and the majority one Z. r^+ and s are samples from Z, such that r^+ is as distant as possible from r^- in terms of FDs, while s is just a random sample from Z without any FD consideration.

The purpose is now to apply classification algorithms on different datasets and compare their performances. The first objective is to see if it is easier to classify between distant datasets, and the comparison is therefore done between r^- versus r^+ and r^- versus s. Classification datasets are constituted only by combining two relations and adding one additional *class* attribute.

	A	B	C	class
r^-	3	3	3	$-$
	7	3	3	$-$
	2	3	2	$-$
	2	3	2	$-$
r^+	1	1	1	$+$
	1	1	6	$+$
	1	4	6	$+$
	1	5	5	$+$

Example 4. *Let's take relations r^- and r^+ from Example 3. They constitute the following classification dataset:*

Using the available datasets, we build two classification models:

- r^- versus r^+
- r^- versus s

Each dataset has then to be divided into training and testing sets, with proportions to be given depending on the experimentation setting. Algorithms are then trained on the training set and their performances evaluated on the testing one. For this preliminary study, the score used to compare classifiers is **accuracy**. This first test relates to one our initial questions, which is whether or not it is easier to classify between distant sets.

For our second experimentation, we come back to the problem of imbalanced datasets. To do so, the same relations can be used but with a slight change in the testing and training sets. Indeed, the philosophy of the data generation presented previously is to emulate such a problem, with r^+ and s being two different undersampling of a bigger relation Z. Therefore we train the models exactly as before, but the testing sets are now different: both models are now evaluated using tuples from r^- and Z, as in a real imbalanced dataset scenario.

5   Experimentations

5.1   Implementation

The algorithms have been implemented using Python 3. All classification algorithms are from the scikit-learn machine learning library [22].

Ten classification algorithms were selected for the experimentations (see [13] for details), with a fixed parametrization as follows:

- K Nearest Neighbors: classification according to the class of surrounding examples. Fixed $k = 3$.
- Decision Tree: learns decision rules and builds a tree. Fixed a maximum depth of 5 for the tree.

- Random Forest: several decision trees on different subsamples of the data. Maximum depth of 5 for 10 trees in total.
- AdaBoost on a decision tree: give different weights to examples, by increasing the weight of misclassified examples.
- Neural Network: fixed two hidden layers with 12 neurons each.
- Naive Bayes: probabilistic model based on Bayes' theorem, with the "naive" assumption that variables are independent.
- Quadratic Discriminant Analysis: finds a quadratic decision surface.
- Linear Support vector machine: classic SVM with linear kernel
- Radial Basis Function (RBF) kernel Support vector machine: RBF kernel.

5.2   Semantic Distance and Classification

The purpose of the first set of experimentations was to study the first conjecture on which this paper is based: it is easier to classify between *distant* sets. The experimental verification of this is simple and follows the explanations given in Sect. 4.

For experimentations, the size of the schema is fixed to $n = 12$. Then:

- $|\mathcal{P}(\mathcal{R})| = 4096$
- $|r^+ \cup r^-| = 4099$

Moreover, sf is fixed to 100: $|Z|$ is therefore around 200 000 tuples. Training test is composed of 80% of a dataset, the remaining 20% are for testing.

The experience was done on ten different instances generated with Algorithms 1 and 2, with each time a new closure system generation and new relations, considering the random components of each algorithm. This is done to make sure any observation is not due to a specific relation, but really a general phenomenon.

Table 1. Accuracy of each classifier for each data generation strategy. Both models are evaluated on their own testing sets.

| Classifier | $ADOM(Z) = ADOM(r^+)$ | | $ADOM(Z) = ADOM(r^+ \cup r^-)$ | |
	r^- vs r^+	r^- vs s	r^- vs r^+	r^- vs s
Nearest Neighbors	0.95	0.87	0.93	0.77
Decision Tree	0.99	0.99	0.99	0.96
Random Forest	0.99	0.99	1.0	0.99
AdaBoost	0.99	0.99	0.99	0.99
Neural Net	0.81	0.72	0.85	0.77
Naive Bayes	0.99	0.99	1.0	0.75
RBF SVM	0.82	0.79	0.77	0.70
Linear SVM	0.62	0.48	0.67	0.47

Table 1 presents the average accuracy score obtained for each algorithm over the ten iterations, when data is generated such that $ADOM(Z) = ADOM(r+)$, so that Z and r^- do not share any common value. Those results are very encouraging, as it appears that classifiers perform better for the distant instances, supporting the conjecture that is easier to classify between distant sets. Indeed, this tendency does not appear to be limited to only one or a few algorithms. Moreover, if the observed difference is anecdotal for some algorithms such as Random Forest and Adaboost, it is pretty important for others. This also opens the way for other paths of research, on how each specific algorithm can be affected by functional dependencies, or how functional dependencies could be integrated in the algorithms themselves to improve their performances. Finally, the active domain on the majority class does not seem to affect our observations, as results are also good with noisy data, and even better in some cases.

5.3   General Imbalanced Dataset Problem

With the good results obtained for the first experimentation, a second one was conducted, closer to the initial imbalanced dataset problem. Indeed, in the end, a classifier should be able to classify correctly on the general dataset: it is trained on a subsample of the majority class, but in the end will be confronted to the real dataset where the classes are imbalanced again. Therefore, in this second experimentation, the training sets stay the same: two balanced datasets, one with two classes voluntarily built as *distant*, and another with a random sample of the majority class against the minority one. But the testing conditions are different, as the objective is now to see if this different training can improve classifier's performances on the imbalanced dataset. The testing set is therefore an imbalanced dataset, with all samples from r^- and Z that have not been used for training.

The conditions are exactly the same as previously with $|\mathcal{R}| = 12$ and $sf = 100$. Results are presented in Table 2. They are less positive than the ones observed in Table 1: the difference between the two models is less pronounced, and seems to be more algorithm-specific.

The results of this experimentations are mitigated, but also encouraging: considering the difficulty and novelty of the considered problem, more work is required, but this first study shows that there is interesting things to investigate. The choice of the values for the synthetic relations is a central problem, and it might be a potential source of improvement. Other types of data could also be used; if letters were used instead of integers, this might bypass the algorithms that rely more or the values.

Table 2. Accuracy of each classifier for each data generation strategy. Both models are evaluated on the same testing set, corresponding to data from an imbalanced datasets, with tuples from r^- and Z.

Classifier	$ADOM(Z) = ADOM(r^+)$		$ADOM(Z) = ADOM(r^+ \cup r^-)$	
	r^- vs r^+	r^- vs s	r^- vs r^+	r^- vs s
Nearest Neighbors	0.70	0.72	0.68	0.71
Decision Tree	0.79	0.81	0.72	0.74
Random Forest	0.95	0.87	0.92	0.86
AdaBoost	0.75	0.78	0.70	0.73
Neural Net	0.66	0.70	0.66	0.77
Naive Bayes	0.83	0.78	0.94	0,75
QDA	0.75	0.89	0.80	0.85
RBF SVM	0.83	0.56	0.77	0.55
Linear SVM	0.99	0.99	0.99	0.99

6 Related Work and Conclusion

The work presented in this paper brings together two problems that do not seem to have been combined before: functional dependencies, a powerful notion in databases on the one hand, and the imbalanced datasets problem, which often occurs in classification, on the other hand. It could impact many application domains, from medical diagnosis to fraud detection. Many overviews on how to handle classification datasets can be found, with various methods relying on diverse computing and mathematical tools. Some solutions work at the data level and resample the data distribution to work on a balanced dataset, with undersampling of the majority class as exposed in this paper, or oversampling of the minority one [7]. Other solutions focus on the algorithms, by trying to adapt them for imbalanced datasets [12]. Finally there are solutions combining both data and algorithmic solutions for imbalanced datasets, such as cost-sensitive approaches [25], as well as boosting algorithms [11]. However some recent works try to address the imbalanced classes problem from a different perspective, like [23]: they use the Mahalanobis distance to create synthetic samples for the majority class.

Functional dependencies have proven to be powerful, in the design of databases [1], or data quality and data cleaning [5], or even to constrain the parameters of type classes in languages such as Haskell [16]. In this paper, we propose to use functional dependencies for the well-known problem of imbalanced datasets. If some results on the classification of distant sets are encouraging, they do not, for now, highlight if functional dependencies could significantly improve the results in imbalanced classification problems. It should however be kept in mind that this is only a first study, that addresses a completely new problem, that is both complex and difficult.

Further experiments are required in future works; studying other data generation strategy, and applying our approach to real data.

It could be argued that on real data, the approach presented could suffer from the lack of existence of functional dependencies in real datasets: however this can be tackled by releasing a bit the constraint of functional dependencies. This is a subject that has already been addressed abundantly in the literature, especially with the concept of fuzzy functional dependencies (FFDs) [15]. The application of this approach to real data will therefore be a logical continuation of this work, but is far from trivial and certainly raises a few combinatorial problems: given a set of functional dependencies, how to select the tuples in the dataset such that they satisfy those FDs, or as much as possible of them ? Moreover, this first contribution opens the way to many more possibilities, as there are presumably other machine learning or data mining problems that could benefit from the use of functional dependencies.

References

1. Abiteboul, S., Hull, R., Vianu, V.: Foundations of Databases: The Logical Level. Addison-Wesley Longman Publishing Co., Inc., Boston (1995)
2. Agrawal, R., Srikant, R., et al.: Fast algorithms for mining association rules. In: Proceedings of 20th International Conference Very Large Data Bases, VLDB, vol. 1215, pp. 487–499 (1994)
3. Armstrong, W.W.: Dependency structures of database relationship. In: Information Processing, pp. 580–583 (1974)
4. Beeri, C., Dowd, M., Fagin, R., Statman, R.: On the structure of armstrong relations for functional dependencies. J. ACM (JACM) **31**(1), 30–46 (1984)
5. Bohannon, P., Fan, W., Geerts, F., Jia, X., Kementsietsidis, A.: Conditional functional dependencies for data cleaning. In: 2007 IEEE 23rd International Conference on Data Engineering, ICDE 2007, pp. 746–755. IEEE (2007)
6. Bonifati, A., Ciucanu, R., Staworko, S.: Interactive join query inference with JIM. Proc. VLDB Endow. **7**(13), 1541–1544 (2014)
7. Chawla, N.V., Bowyer, K.W., Hall, L.O., Kegelmeyer, W.P.: SMOTE: synthetic minority over-sampling technique. J. Artif. Intell. Res. **16**, 321–357 (2002)
8. Chiang, F., Miller, R.J.: Discovering data quality rules. Proc. VLDB Endow. **1**(1), 1166–1177 (2008)
9. Cumin, J., Petit, J.-M., Scuturici, V.-M., Surdu, S.: Data exploration with SQL using machine learning techniques. In: International Conference on Extending Database Technology-EDBT (2017)
10. Dimitriadou, K., Papaemmanouil, O., Diao, Y.: AIDE: an active learning-based approach for interactive data exploration. IEEE Trans. Knowl. Data Eng. **28**(11), 2842–2856 (2016)
11. Freund, Y., Schapire, R.E.: A decision-theoretic generalization of on-line learning and an application to boosting. J. Comput. Syst. Sci. **55**(1), 119–139 (1997)
12. Ganganwar, V.: An overview of classification algorithms for imbalanced datasets. Int. J. Emerg. Technol. Adv. Eng. **2**(4), 42–47 (2012)
13. Han, J., Pei, J., Kamber, M.: Data Mining: Concepts and techniques. Elsevier, Amsterdam (2011)

14. Imielinski, T., Mannila, H.: A database perspective on knowledge discovery. Commun. ACM **39**(11), 58–64 (1996)

15. Jekov, L., Cordero, P., Enciso, M.: Fuzzy functional dependencies. Fuzzy Sets Syst. **317**(C), 88–120 (2017)

16. Jones, M.P.: Type classes with functional dependencies. In: Smolka, G. (ed.) ESOP 2000. LNCS, vol. 1782, pp. 230–244. Springer, Heidelberg (2000). https://doi.org/10.1007/3-540-46425-5_15

17. Katona, G.O.H., Keszler, A., Sali, A.: On the distance of databases. In: Link, S., Prade, H. (eds.) FoIKS 2010. LNCS, vol. 5956, pp. 76–93. Springer, Heidelberg (2010). https://doi.org/10.1007/978-3-642-11829-6_8

18. Kotsiantis, S., Kanellopoulos, D., Pintelas, P., et al.: Handling imbalanced datasets: a review. GESTS Int. Trans. Comput. Sci. Eng. **30**(1), 25–36 (2006)

19. Kotsiantis, S.B., Zaharakis, I., Pintelas, P.: Supervised machine learning: a review of classification techniques. In: Emerging Artificial Intelligence Applications in Computer Engineering, vol. 160, pp. 3–24 (2007)

20. Levene, M., Loizou, G.: A Guided Tour of Relational Databases and Beyond. Springer, Heidelberg (2012)

21. Müller, H., Freytag, J.-C., Leser, U.: Describing differences between databases. In: Proceedings of the 15th ACM International Conference on Information and Knowledge Management, pp. 612–621. ACM (2006)

22. Pedregosa, F., et al.: Scikit-learn: machine learning in Python. J. Mach. Learn. Res. **12**, 2825–2830 (2011)

23. Sharma, S., Bellinger, C., Krawczyk, B., Zaiane, O., Japkowicz, N.: Synthetic oversampling with the majority class: a new perspective on handling extreme imbalance (2018)

24. Shen, Y., Chakrabarti, K., Chaudhuri, S., Ding, B., Novik, L.: Discovering queries based on example tuples. In: Proceedings of the 2014 ACM SIGMOD International Conference on Management of Data, pp. 493–504. ACM (2014)

25. Zadrozny, B., Langford, J., Abe, N.: Cost-sensitive learning by cost-proportionate example weighting. In: 2003 Third IEEE International Conference on Data Mining, ICDM 2003, pp. 435–442. IEEE (2003)

26. Zou, B., Ma, X., Kemme, B., Newton, G., Precup, D.: Data mining using relational database management systems. In: Ng, W.-K., Kitsuregawa, M., Li, J., Chang, K. (eds.) PAKDD 2006. LNCS (LNAI), vol. 3918, pp. 657–667. Springer, Heidelberg (2006). https://doi.org/10.1007/11731139_75

Author Index

Printed in the United States
By Bookmasters